About the author

Judith Richter, born 1954 in Germany, is a pharmacist, tropical medical biologist and development sociologist by training, and a health researcher and activist by professional choice. She has lived and worked in a number of countries. She spent over four years in Thailand, working first as a lecturer in community pharmacy at Khon Kaen University and later as a researcher and international information coordinator of a Bangkok-based consumer protection group, the Drug Information for Action Centre. Since she was invited as a consumer representative to a major symposium on immuno-contraceptives held by WHO's Special Programme on Research, Development and Research Training in Human Reproduction, in 1989, she has been interested in the medical, social and ethical aspects of this research line. She has been writing and publicly speaking on the development of immuno-contraceptives for the past four years. Presently, she is a scholar at the Centre for Ethics in the Sciences and Humanities, at Tübingen University, Germany.

Vaccination against Pregnancy
Miracle or menace?

JUDITH RICHTER

MELBOURNE

Zed Books
LONDON AND NEW JERSEY

Vaccination against Pregnancy: Miracle or menace? was first published in Australia and New Zealand by Spinifex Press, 504 Queensbury Street, North Melbourne, Victoria, 3003, Australia, and in the rest of the world by Zed Books Ltd, 7 Cynthia Street, London N1 9JF, UK, and 165 First Avenue, Atlantic Highlands, New Jersey 07716, USA, in 1996.

This present book is based upon a booklet co-published by BUKO Pharma-Kampagne and Health Action International Europe in 1993.

Cover designed by Andrew Corbett.
Set in Monotype Ehrhardt by Ewan Smith.
Printed and bound in the United Kingdom by Biddles Ltd, Guildford and King's Lynn.

A catalogue record for this book is available from the British Library.

US CIP data is available from the Library of Congress.

Zed Books
ISBN 1 85649 281 8 cased
ISBN 1 85649 282 6 limp

Spinifex Press
ISBN 1 875559 57 4

Australian CIP data is available from the National Library of Australia.

Contents

Acknowledgements

Numerous people throughout the world have contributed their thoughts, information and support to this book: health, reproductive rights and development activists; health professionals and scientists; students of medicine, women's, and development studies; and officials of governmental and international organizations. Though I can thank only a few by name I do hope all others feel included too.

The idea for this publication arose during a discussion with Gudrun Hennke (BUKO Pharma-Kampagne) and Anna Sax (Declaration of Berne). I am very much obliged to them for their encouragement and support, without which I could not have undertaken this task.

Many people kindly helped me with my research. Bernhard Maier and Ulrich Schaible, both formerly of the Max-Planck Institute for Immunebiology in Freiburg and now respectively at Virginia University, Norfolk, USA, and Washington University, St Louis, patiently answered many questions on the intricacies of the immune system. Ulrike Schaz and Ingrid Schneider from FINRRAGE Germany provided me with much valuable material and comments, as did Annette Will of the BUKO Pharma-Kampagne, Nicoliev Wieringa and Anita Hardon of the Women's Health Action Foundation (WHAF) in Amsterdam, Chayanika Shah and Swatija Paranjape of the Forum for Women's Health in Bombay, Kalpana Metha of Saheli Women's Resource Centre in New Delhi, Barbara Mintzes of Health Action International (HAI-Europe), and Jane Cottingham of the World Health Organization in Geneva, Laurel Guymer, Shree Mulay and Sumati Nair. Beatrijs Stemerding of the coordination office of the Women's Global Network for Reproductive Rights did a tremendous job in collecting, digesting and redistributing information on the worldwide campaign activities. Numerous others made the effort to send it to her.

Many people reviewed various drafts of the 1993 report and/or this book version of *Vaccination against Pregnancy*: Niru Acharya, Anita Hardon, Andrew Herxheimer, Catherine Hodgkin, Barbara Mintzes, Ingrid Molema, Denise Noel-Debique, Sabine Paul, Grace Samson, Faye Schrater, Molly Thomas, Ellen t'Hoen, Beatrijs Stemerding, Nicolien Wieringa, Annette Will and Thea Wuest-Zaremba.

I thank all of the above for freely sharing their ideas. The responsibility for any shortcomings, however, is wholly mine.

Annette Will and Barbara Mintzes made considerable efforts in editing the first version of this work. They also contributed greatly to its content. Sarah Sexton of the *Ecologist* edited this book version and gave most stimulating criticisms and comments and contributions. It has been a pleasure to work with her. Louise Murray of Zed Books gave most patient and encouraging support. Very special thanks go to all of them.

Funding from BUKO Pharma-Kampagne, Health Action International-Europe, Mama Cash, Terre des Hommes Germany, Action Solidarity World, and the Declaration of Berne, which made the 1993 version of this publication possible, is gratefully acknowledged. Many thanks to the *Ecologist* for their substantial contribution towards the editing costs. Without a most generous loan from my mother, Edith Richter, I would not have been able to take the time to research and write either the first or this second version of *Vaccination against Pregnancy*. I express my deepest gratitude to her.

Final thanks must go to: Jörg Schaaber from the BUKO Pharma-Kampagne and Phil Armstrong for doing the drawings; Henri Robbemond, Ton Rimmelzwan, and Ank van de Bergh for helping in many desperate situations with my ever-ailing computer; and to my friends Loes Keysers, Than-Dam Truong, Cisca Beckers, Ysabel Perez, Sabine Häusler, Mila Avramovic, Jagoda Paukovic, Azza and P. L. Karam-de Silva, Linda McPhee, Michaela and Gabriel Richter and Eva Hummel-Richter for keeping my body and soul together through numerous delicious meals and uplifting chats.

Abbreviations

AIDS	acquired immune deficiency virus
CONRAD	Contraceptive Research and Development Program, Norfolk, USA
CTP	carboxyterminal peptide
FDA	Food and Drug Administration (of the USA)
FINRRAGE	Feminist International Network of Resistance to Reproductve and Genetic Engineering
FSH	follicle-stimulating hormone
GnRH	gonadotropin-releasing hormone
hCG	human chorionic gonadotropin
HAI	Health Action International
HIV	human immunodeficiency virus
HRP	The Special Programme of Research, Development and Research Training in Human Reproduction of the WHO, commonly known as the Human Reproduction Programme
IUD	intra-uterine device
LH	luteinizing hormone
NICHD	National Institute for Child Health and Development, Bethesda, USA
NIH	National Institutes of Health, Bethesda, USA
NII	National Institute of Immunology, New Delhi, India
oLH	ovine LH (sheep LH)
PID	pelvic inflammatory disease
STD	sexually transmitted disease
TSH	thyroid-stimulating hormone
TT	tetanus toxoid
USAID	United States Agency for International Development
VILCI	Vaccine for Induced Local Cell-mediated Immunity
WHAF	Women's Health Action Foundation
WHO	World Health Organization
WGNRR	Women's Global Network for Reproductive Rights

To my mother
who taught me the value of social justice and
allowed me to become the person I am

Introduction

The prospect of regulating fertility by manipulating immune mechanisms has hitherto been a pious hope linking the aspirations of family planners and reproductive immunologists. Now at last the rational and practical application of this exciting yet frustrating tract of science is at hand. (Warren R. Jones, principal investigator of the WHO's Human Reproduction Programme's Phase I human trial, 1982)[1]

Years ago, when I was myself working in endocrinological research, vaccination ideas like this were raised and promptly dismissed as unethical and dangerous; I do not think that the balance of the argument has changed, except that the threat has come closer, and people are now actually being exposed. (Graham N. Dukes, editor, *International Journal of Risks and Safety in Medicines*, 1993)[2]

Since the 1970s, a number of medical research institutions have been developing a totally new class of birth control methods: immuno-contraceptives, also known as antifertility 'vaccines'. The stated aim of the research is to develop a birth control method that would act for one to two years, have no adverse pharmacological effects, and be easy to administer, cheap to produce, and highly acceptable to individuals because of the general popularity and acceptance of anti-disease immunization.

Although still under development, antifertility 'vaccines' are being portrayed in the scientific and general media as a breakthrough in contraceptive research. According to the World Health Organization's research coordinator, David Griffin, 'the vaccine may prove as important a development in birth control as the contraceptive pill' (WHO 1986). The magazine of the United Nations Development Programme, *Choices*, reported that 'if perfected and mass distributed, the vaccine could have enormous implications, especially for women in the developing world' (Barriklow 1993:28).

However, some scientists, women's rights groups and health activists have questioned the reliability and safety of immuno-contraceptives and raised concerns about the contraceptives' potential to be used coercively, about the way in which clinical trials are being carried out, and about the general conceptual framework underlying the development of these methods. *Vaccination against Pregnancy* has been written as a contribution towards the informed public debate about immuno-contraceptives which should take place before they are approved for the market and their distribution and availability become a fact of life.

This study attempts to foresee and assess the possible undesirable effects of immuno-contraceptive research, particularly for women because women are considered to be the main beneficiaries – as well as the main victims – of modern birth control methods. The study will consider the societal forces that have hitherto shaped contraceptive development, and the influence that various interest groups have had on the range of contraceptives currently available – or not available – to women and to men.

Historical dynamics of contraceptive research

Modern contraceptive development has a distinct history compared to that of other industrially manufactured pharmaceutical drugs or devices. A major difference between them has been that the decision over which birth control method to develop (or not develop) has been determined primarily by the intersecting interests of various population institutions and the scientific community, rather than by the profit motives of the pharmaceutical industry – or by concern for women's health, safety and well-being.[3]

Throughout the ages and in many cultures, women have wanted to have greater influence over their fertility. Despite popular perceptions to the contrary, their wishes have in fact had only a marginal influence on the direction of twentieth-century contraceptive research. Until the 1960s, mainstream scientists and the pharmaceutical industry were reluctant to engage in contraceptive research because they did not want to confront strong social taboos, backed by powerful interest groups, against women's sexual freedom and reproductive self-determination.

Although the pharmaceutical industry and mainstream scientists had had the basic knowledge of how to develop hormonal contraceptives since the 1920s (Oudshoorn 1994:97), it was left to Gregory Pincus, a reproductive scientist who carried out contract research at a private foundation in Massachusetts, to develop the first oral hormonal contraceptive pill in the late 1950s.

Pincus had been asked to do this by Margaret Sanger of the Planned Parenthood Federation. Sanger was looking for an oral contraceptive that would be easy to use because she believed that poorer people were less adept at using what were then the most widely used contraceptive methods or practices, namely the diaphragm, vaginal douche, condom, rhythm method or withdrawal. Like many of her contemporaries, Sanger was worried about 'race suicide' in the USA, fearing that poor, particularly black, people would 'outbreed' white, middle-class Americans. Although initially women-centred, her perspective had gradually shifted towards a eugenic point of view as she declared in 1919, 'More children from the fit, less from the unfit – that is the chief issue of birth control' (in Gordon 1990: 277).[4]

Feminist research has shown that the rise and endorsement of population control ideology among certain interest groups in the late 1950s and early 1960s was a decisive factor in turning contraceptive development from an endeavour that medical scientists shied away from into a socially acceptable, even desirable, one (Clarke 1995; Hartmann 1995).

The various negotiations between different interest groups that determined the nature of modern contraceptives, however, took place prior to this, primarily between 1925 and 1945 in countries such as the USA and Germany. The main actors were advocates of birth control (including feminists who wanted reproductive self-determination for women, medical professionals who wished to lower maternal mortality, eugenicists and neo-Malthusians) and reproductive scientists (Clarke 1995).

During this time, there was a shift from regarding birth control as a means of individual self-determination to perceiving it as a tool of population control. This shift was paralleled by a move in contraceptive research away from 'simple', user-controlled means of contraception such as the diaphragm and chemical spermicides towards 'scientific' means of contraception based on 'basic reproductive science' (Clarke 1995). During this period, research started into hormonal contraception, intrauterine devices (IUDs), immunization against sperm and sterilization by radiation.

US corporate foundations such as the Rockefeller Foundation and the Ford Foundation have been instrumental in creating institutions such as the Population Council and networks of scientists to engage in contraceptive development. In the 1960s, their lobbying of US population control proponents resulted in a substantial flow of public funds for contraceptive research and distribution (LaCheen 1986; Heim and Schaz 1994). At this time, however, the majority of contraceptive research funds came from pharmaceutical companies attracted by the potential of an enormous market for products which women throughout the world would use over their thirty or so fertile years.

But during the 1970s, most large US pharmaceutical companies started to pull out of research and development into new contraceptives. This was in part the result of protests from women's and consumer groups worldwide over the inadequate safety testing of contraceptives that had already come on to the market, some of which had become the subject of litigation. Various liability suits against the contraceptive manufacturers for misleading safety claims for the first generation of (high-dose) Pills and IUDs (in particular the Dalkon Shield) had resulted in tighter standards for medical research in general prior to a product being registered or licensed. This caused research and development costs to increase. Some women's and Third World solidarity groups also opposed the introduction of less-safe contraceptive products, in particular, the injectable hormonal contraceptives Depo Provera® and Net-En®. As a result, large pharmaceutical companies decided that contraceptives were too controversial and yielded too little profit compared to other products and thus pulled out of contraceptive research, manufacturing and marketing (Mastroianni et al. 1990:61).

Today, there is in effect a dual research agenda into birth control methods: larger pharmaceutical companies tend to leave the financial investment and risks of new contraceptive research to governments, national and international organizations such as the Population Council and the Human Reproduction Programme (HRP)[5] of the World Health Organization (WHO) and small entrepreneurial firms, while concentrating themselves on improving existing hormonal contraceptives for customers who can afford to pay for them (Gelijns 1991). Such companies have helped to perfect and market new contraceptives only if they are satisfied that there will be sufficient profit potential and that the research institutions have fostered public acceptability for these methods.

For example, Norplant® – six matchstick-sized rods that are inserted into a woman's upper arm to release hormones over a five-year period – was developed primarily by the Population Council, an organization described by its vice-president, C. Wayne Bardin, as one that 'applies science and technologies to the solution of population problems, mainly in developing countries' (Bardin 1993). The council subsequently contracted with a small Finnish company, Leiras, to finalize Norplant®'s development, to register it for use in Finland in 1983 and to introduce it in Finland, Sweden and several Third World countries from 1984 onwards. The large US pharmaceutical corporation Wyeth, which holds the patent on the synthetic hormone used in Norplant®, applied for the implant's approval on the US market only several years later – approval was granted in 1990 (Hartmann 1995:208). The development and market introduction of Norplant® was funded by six private foundations, the US and Finnish governments, and Leiras and Wyeth (Bardin 1993).

Non-profit investment in contraceptive research has focused on new contraceptives designed primarily for use in subsidized programmes aimed at reducing population growth rates in Third World countries. The focus has been to develop methods that are highly effective in preventing pregnancy, that act for a long period of time ranging from several months to several years, and that have a low risk of 'user failure'. (By 'user failure', contraceptive developers mean the likelihood that the contraceptive method fails to prevent pregnancy because the woman or man does not use the method 'properly'.) Less of a priority have been issues of safety and women's self-determination, in particular, the question as to whether or not a woman can stop the effect of a contraceptive any time she wishes.

Research into barrier methods (such as condoms or diaphragms), improvement of natural family planning methods and male contraceptives, and research into traditional methods and practices of fertility control have received only a fraction of worldwide contraceptive funding in the last thirty-five years.

Antifertility 'vaccines' – ideal for whom?[6]

The idea of inducing infertility via the immune system is not a novel one. A connection between the actions of the immune and reproductive systems has been known since the end of the nineteenth century when experiments conducted by three leading immunologists, the Austrian Karl Landsteiner and the Russians Elie Metchnikoff and Sergei Metalnikoff, indicated that injection of sperm or testes extracts into animals could induce the animals to form antibodies against sperm (in Katsh 1959:947).

While this knowledge prompted some scientists to carry out research into the potential immunological causes of existing infertility, others began to advocate the development of immunological methods to induce it. Warren Jones has described how a 'wave of Malthusian fervour in the 1920s ... led to the performance of a unique series of studies of immunological contraceptives in man'. Actually, these studies, at least twelve of them conducted between 1920 and 1934, were carried out in women (Joel 1971, quoted in Jones 1982:9). Ease of administration and low costs were seen as the major advantages of immunological contraception:

> Think how wonderful it would be if one could immunize a patient by a simple hypodermic injection once every six months, just as we today immunize children against diphtheria. It will indeed be a new and wonderful era in the practice of preventive gynecology. (Daniels 1931, quoted in Clarke 1995)

> Devices are all very nice for those who can afford them. The poor people with whom we are really concerned in this [Depression] recovery program cannot afford them ... It is quite necessary to be concerned with something that can be applied much more cheaply. Spermatotoxins [anti-sperm 'vaccines'] ... are one of the methods. (McCartney 1934, quoted in Clarke 1995)

However, a discouraging evaluation report by the US National Committee on Maternal Health in the late 1930s regarding the efficacy of immuno-contraceptives put an end to this line of research. The report concluded:

> When one compares ... the fertility of the injected animals with the controls, it appears that ... injection of live sperms reduces slightly the fertility of the

recipients, but the reduction is neither of significant degree nor of practical importance. (Eastman, Guttmacher and Stewart 1939, quoted in Clarke 1995)

Two decades later, Seymour Katsh of the University of Colorado Medical Center in Denver, USA, undertook a review, funded by the Population Council, of these first studies in order to 'stimulate intensive experimentation and clinical investigation in this broad area of the control of infertility and fertility by immunological means'. Katsh considered that the 'core' of an urgently needed 'control of populations' was the availability of an 'effective and reliable method ... for inducing infertility (temporary or permanent) in both males and females which [would] meet with the approval of the greatest number of people' (Katsh 1959:946).

Despite Katsh's hopes, however, the time was still not yet ripe for further research into immuno-contraceptives. It was not until the late 1960s and early 1970s that advances in immunology and a favourable funding climate caused by increased worries about 'population' prompted immunologists to take up the interrupted endeavour. 'Immunization as a prophylactic measure is now so widely accepted that it has been suggested that one method of fertility control which would have wide appeal as well as great ease of service delivery would be an antifertility vaccine', the Human Reproduction Programme of the WHO stated in 1978 as a second wave of research and development into immuno-contraceptives got under way (HRP 1978:360).

The aspirations of these second-wave developers go far beyond inducing immune reactions in women against men's sperm. Research is being conducted into the inducing of immune reactions against reproductive and pregnancy hormones and against egg cells and sperm cells.

As such, immuno-contraceptives are more than just a potential addition to the existing array of contraceptives. They constitute a whole new class of contraceptives, the benefits and risks of which depend significantly on the target of the immune reaction, that is, whether they attempt to induce an immune reaction against reproductive- or pregnancy-related hormones, or against egg cells or sperm cells. In theory, immuno-contraceptives could be developed for use by both men and women, but so far most of the research has been directed towards immuno-contraceptives to act in women's bodies.

As was the case with the Pill, IUD and Norplant®, the prime mover behind the development of immuno-contraceptives is not the pharmaceutical industry. By the mid-1990s, five research centres were coordinating the bulk of research:

— the National Institute of Immunology (NII), New Delhi, India;
— the Special Programme of Research, Development and Research

Training in Human Reproduction (HRP) of the WHO, Geneva, Switzerland;
— the Population Council, New York, USA;
— the Contraceptive Research and Development Program (CONRAD), Norfolk, USA;
— the National Institute of Child Health and Development of the National Institutes of Health (NICHD/NIH), Bethesda, USA.

Immuno-contraceptives are also being researched by several smaller university research teams, including those working under N. Raghuveer Moudgal at the Indian Institute for Science in Bangalore, India; John Aitken at the Reproductive Biology Unit at Edinburgh University, UK; John C. Herr, Eastern Virginia Medical School, USA; and Dominique Bellet at the Institut Gustave Roussy, France.[7]

Funding for this research has come from a variety of public sources including the governments of the United States, India, Canada, the UK, Norway and Germany, as well as from corporate philanthrophic sources, in particular the Rockefeller Foundation.[8]

Although research on immuno-contraceptives has not yet progressed beyond the second of three human trial stages, immunologist Avrion Mitchison believes that 'it is now generally accepted that vaccines will come to be used for the control of fertility' (Mitchison 1990).

Vaccination against Pregnancy examines whether the optimism of the research community is justified. Are antifertility 'vaccines' an acceptable birth control option – and if so, acceptable for whom?

As a health professional and consumer activist, I became concerned about immuno-contraceptives in 1989 when I was invited as one of two consumer representatives to attend a symposium held by the WHO's HRP addressing the medical, social, legal and political aspects of these contraceptives. I was subsequently invited by the HRP to participate in two further meetings, one in August 1992 between researchers and women's health advocates, and another in June 1994 between the HRP's ethical committee and women's health advocates (see Appendix 1; Ada and Griffin 1991a; HRP 1993; HRP and SERG 1994). My observations and experiences at these three meetings combined with many discussions with scientists and health professionals from different disciplines and with women's and health activists from all over the world as well as the study of scientific reports form the basis for this analysis.

Chapter 1 provides an overview of the different types of immuno-contraceptives currently being researched and developed and summarizes how they are intended to operate so as to induce the body's immune responses to interfere with reproduction.

Chapter 2 describes the fundamental problems researchers and developers face if they are to ensure that immuno-contraceptives are to be effective, reversible and safe, both for users and for any future offspring – the minimum requirements for any new contraceptive. It explains why solutions to various problems developers have encountered cannot easily be extrapolated either from knowledge about anti-disease vaccines or from naturally occurring immune-mediated infertility.

Chapter 3 describes and anticipates the biomedical characteristics of immuno-contraceptives and their potential risks from a user-centred perspective.

Chapter 4 assesses the potential for abuse of immuno-contraceptives (that is, the potential for the method to be used coercively or without the fully informed consent of the person receiving the contraceptive), a potential for abuse that arises not only from the possibility of 'over-zealous' contraceptive providers but also from the actual design of the contraceptive itself. Both these assessments attempt to consider the real-life situation of those who will probably be the major 'target group' of immuno-contraceptives, namely impoverished, relatively powerless women who often have little access to health care. A critical question is whether immunological birth control methods can ever fulfil the requirement of being reliable and safe individual contraceptives.

Chapter 5 analyses the conceptual framework that has guided the development of antifertility 'vaccines' and investigates whether the five major advantages attributed by the research community to immuno-contraceptives are advantages from the perspective of future users.

Chapter 6 provides a glimpse into immuno-contraceptive research practice. According to internationally accepted codes of conduct in medical research, trials on humans are only justified if the new therapeutic drug or contraceptive product being trialled offers an advantage over existing products; the risks for the individual trial participants must be minimized; and researchers must ensure that participation is voluntary and informed.

Chapter 7 describes the emergence and building of a worldwide campaign against the continued development of immuno-contraceptives. By June 1995, over 430 groups and organizations from nearly forty countries were calling for a stop to this particular line of research and for a radical reorientation of contraceptive research to put people's integrity, health and well-being at its centre rather than population control. The chapter analyses the reactions of the research and funding community to the concerns the coalition has raised, and assesses the likelihood of influencing the research agenda by summarizing some of the developments since the campaign was launched.

Throughout *Vaccination against Pregnancy* I tend to refer to these new birth control methods as 'immuno-contraceptives' rather than 'vaccines' because they differ in significant ways from vaccines against diseases.[9] Vaccines stimulate normal immune responses against micro-organisms to ward off diseases, whereas immuno-contraceptives induce a disorder of the immune system, namely immunological infertility. As pregnancy is a natural and healthy process which is neither a disease nor an epidemic and as the foetus is not a harmful germ invading the body, preventing a pregnancy is profoundly different from preventing disease (and so are the criteria underlying the benefit/risk assessments of anti-disease vaccines and immune-mediated contraceptives). In choosing the term 'immunological contraceptives', I have not made a clear distinction between those immunological methods that act either before or after conception because I am above all concerned about people's self-determination and well-being.

Notes

1. Jones 1982:8.
2. In Wieringa 1994:2.
3. For more information, see LaCheen 1986, Hartmann 1995, Gordon 1990, Heim and Schaz 1994; Clarke 1995.
4. Incidentally, when the Pill became available in the early 1960s, it was white, middle-class women who took it up with enthusiasm. Subsequently, many mainstream accounts of contraception have hailed the availability of the Pill as a milestone in women's emancipation, confounding the unintended and unforeseen outcome of a technology with the intentions behind its development. See e.g. Reed 1983.
5. The Special Programme of Research, Development and Research Training in Human Reproduction is commonly referred to as the Human Reproduction Programme or HRP. It was established as a Special Programme of the World Health Organization in 1972. Since 1988 when the United Nations Development Programme (UNDP), the United Nations Fund for Population Activities (UNFPA) and the World Bank pledged regular funding for it, the programme has been called the UNDP/UNFPA/WHO and World Bank Special Programme of Research, Development and Research Training in Human Reproduction. For ease I refer to it throughout as HRP.
6. The heading is borrowed from a flyer by the Buko-Pharma-Kampagne and FINRRAGE, published in 1994 and entitled *Anti-fertility vaccines: the ideal contraceptive - for whom?*
7. In addition to these, there are many small teams worldwide, many of which are concentrated in the USA because of available funding from USAID and the US National Institutes of Health.
8. Between 1985 and 1995 10 per cent of the public funding for contraceptive research went towards the development of immuno-contraceptives (Spieler 1992).
9. In response to my suggestion that the metaphorical term 'antifertility vaccine' should be abandoned, Indian women's rights activists Swatija Paranjape and Chayanika Shah commented to me in 1993: 'We continue to call it the "antifertility vaccine" and not "immunological contraceptives". We feel that the basic assumption behind the development of this contraceptive is an understanding of fertility as disease – a communicable one at that! It is considered to

be an epidemic in the context of the poor, marginalized women all over the world. The name that [the researchers] have given highlights their mentality in producing it and we who are against developing any such method of contraceptive should not be giving them a term under whose garb their real motives can be hidden.' They acknowledged the need, however, always to point out that these 'vaccines' are in fact contraceptives using the immune system – even if the term 'immunological contraceptives' may be difficult to translate into local languages. Personally, I am now using the term 'immuno-contraceptives' when referring to the biomedical characteristics and risks of these products, and antifertility 'vaccines' (between quotation marks) when referring to the risk of abuse.

I

Immuno-contraceptives: a new class of birth control methods

No method of regulating fertility has ever before rested on immunological principles, nor has any vaccine ever been directed towards the inhibition of a 'self-like' component or secretion. (Jeff Spieler, Senior Biomedical Research Adviser, Office of Population, USAID, 1987)[1]

The immune system

Most people think of the immune system as the body's means of warding off illness and disease so as to keep us healthy. It is so complex, however, that its workings and its connections with human reproduction are still not fully understood. To describe some of these complexities and functions and the attempts being made to turn the immune system against reproductive components, I use a simplified, somewhat mechanistic, molecular model. This model, which is just one way of describing the immune system, is one that is rapidly evolving. It is important to remember that the immune system is interconnected with many other 'systems' and is responsive to many other influences besides germs or micro-organisms.[2] Ultimately at issue, however, are not molecules or systems but people, their fertility, their health and that of their future children.

Our immune system is a collection of interacting cells, molecules and tissues. Most of us know about the immune system from its role in countering the effects of micro-organisms (such as viruses, bacteria and fungi) as well as parasites which can cause illness and disease. The immune system's functioning is based on two features: its ability to respond specifically to each type of micro-organism, and its 'memory'. Once the immune system has encountered a specific micro-organism, it 'remembers' this micro-organism and generates a more efficient immune response every time it subsequently encounters it.

When an infectious organism enters the human body, the immune system 'recognizes' the organism as 'foreign' to the body and tries to eliminate it. The first time the immune system encounters a specific micro-organism, its response – the *primary immune response* – is relatively slow and inefficient. A person usually becomes ill, recovering as the immune response against the disease-causing micro-organism takes hold

– or, depending on the micro-organism and the status of the immune system, not recovering and even dying. This primary response triggers the formation of *memory cells* which 'remember' the specific micro-organism. Subsequent infections with the same micro-organism elicit a *secondary immune response*, a faster and more vigorous reaction against the organism. A secondary immune response usually either prevents the outbreak of illness or disease or greatly reduces its severity. A person may not even be aware that he or she has been infected. This memory is refreshed each time the immune system encounters the disease-causing agent.

For example, once people who contract measles, usually in childhood, have succumbed to the infection and recovered, they are usually 'immune' to subsequent measles infections. The next time the measles viruses enter the body, their spread is prevented by the immune cells which have learned to recognize and react against them.

A specific type of white blood cells, the *lymphocytes*, are responsible for both the specific memory and the production of specific antibodies and immune cells which neutralize or destroy the infectious agent.

Traditional vaccines aim to build up the immune resistance to a specific harmful disease without the person having to succumb to it. This is done by giving the person a version of the disease-causing agent or toxin (its poisonous secretion) which has been altered so as to stimulate the immune response but not to cause the disease. For example, if someone contracts the disease diphtheria, they can die because diphtheria micro-organisms secrete a potentially fatal toxin. However, if the person has been vaccinated with an altered, non-toxic, version of the diphtheria toxin, called a 'toxoid', the immune system is primed and will react more quickly and vigorously to the diphtheria-causing toxin if it encounters it.

The immune system not only 'recognizes' foreign micro-organisms but also recognizes cellular and molecular components of our bodies. It was discovered relatively recently that many people possess a number of *auto-antibodies* – antibodies that recognize and could potentially harm their own body components – but are in fact healthy. It seems that the immune system can differentiate between 'foreign' ('non-self') and 'self' *antigens* (materials, either foreign organisms or body components, which the immune system recognizes) and learns to eliminate or neutralize the foreign ones but not to react against self-antigens. Although 'self' structures may be antigens, they do not usually act as *immunogens*, that is, they do not induce an immune response to eliminate or neutralize them. This protection of a person's own body constituents against attack by the immune system is known as *self-tolerance*.

How self-tolerance functions is not fully understood, but it is known that the immune system's learning process starts at the foetal stage. If self-tolerance fails to function properly, *auto-immune diseases* may occur, that is, diseases in which the immune system acts against specific body components as if they were 'foreign' and tries to eliminate them from the body.

For example, when the immune system reacts against the lining of the joints, rheumatoid arthritis (a recurrent inflammatory disease which may result in erosion, destruction and deformity of the joints) develops.[3] When the immune system reacts against insulin-producing cells in the pancreas, juvenile diabetes may result. Some auto-immune diseases may go into remission spontaneously; many do not. Although the symptoms of some auto-immune diseases can be treated, the underlying specific immune reactions against particular body constituents cannot yet be altered or stopped. In extreme cases, pharmaceutical drugs are prescribed which suppress immune responses in general, interfering not only with the specific harmful auto-immune reaction but also with the body's immune reactions against disease-causing agents.

Links between immune and reproductive systems

If the immune system reacts against hormones, cells or other body secretions that are indispensable for successful human reproduction, a particular type of immune disorder may result, namely infertility. It is known, for example, that infertility in women is sometimes caused by a woman's body generating immune reactions against sperm (although many women who do have antibodies against sperm are not infertile).

Immune factors are thought to play a role in many early miscarriages. The most well-known example of an immune-mediated miscarriage is that of a foetus with the rhesus-positive blood type of the father which is rejected by the body of a rhesus-negative mother. Usually, the first pregnancy has proceeded normally, but the mother's immune system came into contact with the rhesus-positive blood of the child during pregnancy or birth, triggering the production in the mother of immune cells that will 'recognize' and 'dissolve' the rhesus-positive blood cells of subsequent foetuses. This immunological dissolution of red blood cells can also cause permanent brain damage in the newborn (fortunately it is possible today to detect and prevent haemolytic disease in the foetus and the newborn).

Immunologists do not know why our immune system tolerates many of the components involved in our reproductive processes. For example, in the conventional model of the immune system it might be considered that a woman's immune system would consider male sperm as 'foreign', and a fertilized egg and a foetus – half of whose genetic material comes from the father – as 'half-foreign'. Successful fertilization and pregnancy appear to be one of the mysteries of life. Some immunologists therefore choose to classify sperm, embryos and foetuses as 'self-like' to indicate that they enjoy immune self-tolerance. It is, of course, a self-evident fact that women's immune systems do not habitually treat healthy sperm or foetuses as 'foreign' – otherwise our species would have become extinct long ago.

Immunological birth control methods

To create or induce immune-mediated infertility, the trick is to make one or more of the components involved in reproduction and early pregnancy appear 'foreign' to the human immune system so that the immune system 'believes' it is a harmful micro-organism to be eliminated. Several components could be altered to achieve this effect.

The birth of a child is the culmination of a complex series of events and processes which began in the bodies of the parents long before conception. Throughout their fertile years, human beings produce reproductive hormones which regulate in an intricate fashion the monthly ripening of women's egg cells and the lining of the womb and the production and maturation of sperm in men's testes. If sexual intercourse takes place around the time of ovulation – the time when a matured egg is released from a woman's ovaries into the Fallopian tubes – one of the millions of sperm may fuse with the egg cell. The fertilized egg cell divides and gradually becomes an early embryo which secretes human chorionic gonadotrophin (hCG), a hormone essential for the establishment and maintenance of early pregnancy.

The immuno-contraceptives currently being researched are designed to disrupt one of three basic reproductive stages:

— the maturation of egg cells and the production of sperm cells;
— fertilization;
— the implantation and development of the early embryo.

To disrupt any of these processes, the immune system has to be induced to produce antibodies that specifically recognize and act against a reproductive component that is indispensable for the successful outcome of each stage – either reproductive hormones that affect the maturation of egg cells and the production of sperm cells; or the egg and sperm cells themselves; or pregnancy-related hormones essential for the early embryo to implant in the uterus and begin to develop. Most of the immuno-contraceptives under development are directed against only part of a reproductive component, such as a small distinct part of a hormone or a specific structure on the surface of a cell that acts as an *antigen*, that is, it is recognized by antibodies of the immune system.

Reproductive hormones To disturb the maturation of egg cells or the

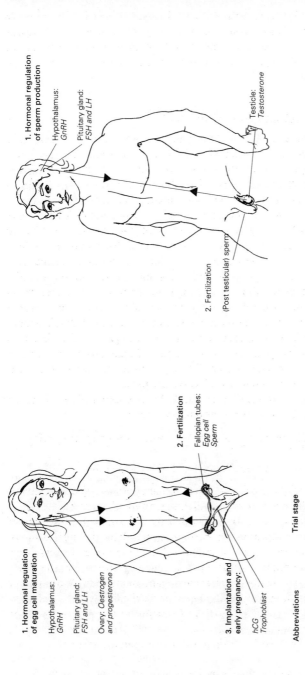

1. Hormonal regulation of sperm production

Hypothalamus: *GnRH*

Pituitary gland: *FSH and LH*

Testicle: *Testosterone*

2. Fertilization
(Post testicular) *sperm*

Trial stage

GnRH: men Phase I
(as anti-cancer product)

FSH: men Phase I

Sperm: men not yet in human trials
(less favoured for men)

1. Hormonal regulation of egg cell maturation

Hypothalamus: *GnRH*

Pituitary gland: *FSH and LH*

Ovary: *Oestrogen and progesterone*

2. Fertilization

Fallopian tubes: *Egg cell Sperm*

3. Implantation and early pregnancy:

hCG Trophoblast

Abbreviations

GnRH: gonadotropin-releasing hormone

FSH: follicle-stimulating hormone

LH: luteinizing hormone

hCG: human chorionic gonadotropin

Trial stage

GnRH: women Phase I
(in breast feeding women)

hCG: women Phase I or II
(2 of 3 prototypes cross-reacting with LH)

Egg cell: women not yet in human trials

Sperm: women not yet in human trials

Figure 1.1 Potential immuno-contraceptives according to reproductive processes and components

production of sperm, one or more of the hormones involved in these processes has to be neutralized:

— GnRH (gonadotropin-releasing hormone)
— FSH (follicle-stimulating hormone)
— LH (luteinizing hormone).

GnRH is secreted by the hypothalamus in the brain and triggers the pituitary gland at the base of the brain to secrete FSH and LH.

In women, FSH and LH act on a woman's ovaries, causing the production of other hormones, oestrogen and progesterone, which in turn affect the maturation and release of an egg from the ovaries during each menstrual cycle. The whole menstrual cycle is thus influenced by a complex feedback between the various hormones released by the ovary, the pituitary gland and the hypothalamus. In men, FSH and LH cause the testes to produce testosterone, which affects sperm production.

Egg and sperm cells   One way of preventing fertilization of an egg by a sperm is to render either the egg cell or sperm cell nonfunctional. Researchers are trying to identify antigenic structures on the surface of the egg cell and sperm cell against which an immune response could be directed. This response would not only act against sperm in men's bodies; some immuno-contraceptives aim to interfere with sperm function in women's bodies shortly before fertilization.

Pregnancy-related hormones and components   The research into immuno-contraceptives that is furthest advanced has been conducted into methods that act against the hormone hCG (human chorionic gonadotrophin), which is essential for the establishment and maintenance of early pregnancy. hCG is secreted in a woman by the early embryo, shortly after the egg has been fertilized. Its effect is to keep a woman's ovary producing progesterone, which in turn causes the lining of the uterus to be maintained in a thickened state in which the embryo can implant itself. If hCG were to be intercepted by antibodies, the progesterone levels would drop – as usually happens if an egg is not fertilized – the lining of the uterus would shed and the woman would have a menstrual-like period.[4]

The trophoblast (part of the early embryo which later forms the placenta) is another potential target antigen being explored. Immuno-contraceptives directed against the trophoblast would have to act so that antibodies would coat it in such a way that it was prevented from implanting in a woman's uterus.

Making reproductive components appear 'foreign' to the immune system
Once a body component that is indispensable for reproduction has been

selected as the target of an immuno-contraceptive, the challenge has been to make the component appear foreign to the immune system. The administration of 'self' antigens alone, either the natural hormone or a fragment of it, does not usually induce a strong enough immune response against them. To make the reproductive antigens appear 'foreign' to the immune system, researchers have linked them to a 'foreign' carrier such as the diphtheria or tetanus toxoid. When the immune system 'encounters' this constructed reproductive antigen–carrier combination, it will react as if the particular reproductive component in the body was a harmful micro-organism to be eliminated.

Even this constructed combination, however, is not usually powerful enough to induce a sufficient immune response that reliably interferes with the reproductive process. Therefore, one or two *adjuvants* – substances that enhance immune responses in general – are added to the formulation. (Besides increasing the desired immune reaction against the targeted reproductive antigen, adjuvants may also enhance ongoing beneficial immune reactions against micro-organisms – and potentially harmful immune reactions such as those causing auto-immune diseases and allergies.)

All of these are usually dissolved in water, but to increase the immune response still further, some of the immuno-contraceptives under development are being devised as a 'slow-release delivery formulation', by being suspended in an oily emulsion. A slower release of the antigen–carrier conjugate from the injection site prolongs the stimulation of the immune system. (More recently, the ingredients have been formulated in some preparations as microspheres in an attempt to increase the duration of efficacy. See p. 34.)

Research to date

Before any biomedical product is approved by a national drug regulatory authority for the market, it has to pass a number of laboratory and animal tests and three stages or phases of human trials. As immuno-contraceptives are a wholly new category of pharmaceuticals, the 1989 HRP Symposium on Assessing the Safety and Efficacy of Vaccines to Regulate Fertility specified the following conditions for research in humans (Report 1991:268–80).

A Phase I (safety) trial should determine the potential risks of the product, in particular, the adverse effects caused by interference with the immune (and, depending on the product, hormonal as well) systems. A secondary aim should be to establish the dose that theoretically may be contraceptive. A Phase I trial should be carried out in sterilized women (to avoid any risk of exposing a foetus in case the immuno-contraceptive has a low efficacy); for a meaningful assessment, around fifty people should participate in the trial. A Phase I trial should take one to two years to complete.

A Phase II (efficacy) trial should aim to assess the efficacy of a specific prototype immuno-contraceptive in preventing pregnancy (and to give further indications of potential risks). The trial requires around two hundred participants of 'proven fertility' (usually women who have had at least two children) and should take two to three years to complete. Before conducting a Phase II trial, it is essential that studies are carried out to determine whether foetal anomalies occur in immunized animals that become pregnant if the immune response is below the contraceptive level.

A Phase III (large-scale) trial should aim to predict 'the efficacy and safety of their use in the general population and in varied service settings'. A Phase III trial should involve around one thousand trial participants and may take four to six years to complete.

The different research teams are now at various stages in developing their specific 'prototype vaccines'. As far as I can determine, to date, around twelve separate trials of immuno-contraceptives have been conducted, involving approximately 430 people in at least seven countries: India, Australia, Sweden, Brazil, Chile, Dominican Republic and Finland (some prototypes have also been tested as anti-cancer products in India, the USA, Austria and Mexico). The majority of these trials have been of the various anti-hCG immuno-contraceptives that act in women. By 1995, more than two decades after the restart of this line of research, only two 'candidate vaccines' had reached Phase II trials: the anti-hCG immuno-contraceptives developed by the NII and by HRP.

Immuno-contraceptives against hCG Different formulations of immuno-contraceptives against the hCG hormone have now been tested in around 360 women. This figure includes women tested with a variety of early hCG prototypes tested between 1974 and 1978 by the NII and the Population Council, and in the more recent trials carried out since the mid-1980s by the NII, HRP and the Population Council.

The Phase I (safety) trial of the Population Council's latest anti-hCG product, completed in 1991, was conducted in three clinics in three different countries: Finland, Chile and the Dominican Republic. The Indian National Institute of Immunology completed its Phase II (efficacy) trial of its latest anti-hCG product in 148 women in three centres in India in August 1993 (Talwar et al. 1994a).

HRP conducted its Phase I (safety) trial at Flinders Medical Centre, Adelaide, in Australia with Warren Jones as the principal medical investigator. At the end of 1993, HRP started its Phase II (efficacy) trial in Sweden intending to enrol 250 women, but suspended it in June

1994 because of the adverse reactions in the majority of the first seven volunteers (HRP 1994b:8).

In addition, the NII has tested an anti-hCG 'vaccine' in at least three lung cancer patients in Mexico. Some lung cancers produce hCG, which seems to promote the cancer's growth (Talwar 1994a:2–3). In the product tested, the hCG antigen was inserted into a live vaccinia virus which replaced the toxoid as a carrier. This way of making hCG appear 'foreign' to the immune system allows for genetic engineering of the anti-hCG 'vaccine' which, according to the researchers, will be cheaper.

Immuno-contraceptives against non-pregnancy-associated reproductive hormones (GnRH, FSH)   Anti-GnRH immuno-contraceptives for women have been given in a preliminary test to twenty Indian women just after giving birth 'to prolong post-partum amenorrhoea' (Talwar et al. 1992b: 513). (Ovulation and menstruation do not usually return for some weeks after childbirth.)

The Population Council and the NII, in cooperation with Julian Frick from the Salzburg General Hospital in Austria, have conducted clinical trials in men with different anti-GnRH preparations. But because the US Food and Drug Administration advised against testing these products as a contraceptive (see Chapter 6), the aim of these trials was to assess whether auto-immunization against GnRH (neutralization of testosterone) stops the growth of prostate cancer (HRP 1993:45–6). Depending on the outcome of these trials, however, researchers plan to test these products as immuno-contraceptives in healthy men. An anti-GnRH product for the treatment of prostate cancer developed by the NICHD has also been approved for clinical trials.

No final reports of these anti-GnRH trials by the NII or the Population Council were available to me. But the NII team has stated of its anti-GnRH product tested in animals that: 'the vaccine impairs fertility of male animals. It has a particularly marked effect on the prostate which atrophies dramatically.' The team also maintains that these effects are reversible: 'With decline of antibodies, testicular functions are restored; the prostate, along with other accessory reproductive organs, regenerates.' As to the product's efficacy in raising the immune response, during 'probing clinical trials' of the product in the prostate cancer patients, the NII team observed that 'the available results indicate that in patients in whom adequate antibody titres [levels] were generated by the [anti-cancer] vaccine, testosterone levels fell to castration levels' (Talwar et al. 1992a:2–3).

Anti-FSH immunization is also being investigated as a means of male fertility control by a team under the head of the Centre for Reproductive Biology and Molecular Endocrinology at the Indian Institute of Science in Bangalore, N. Rhaguveer Moudgal. In early 1995 the team completed the Phase I (safety) trial of anti–FSH auto–immunization in men (Talwar 1994b:699) but the results were not known at the time of going to press.

Immuno–contraceptives against egg and sperm cells Most anti–sperm and anti–egg immuno–contraceptives are still at the laboratory stage, although some have been tested in various animals. Research on this type of immuno–contraceptive is mainly carried out in projects supported by the USAID-funded Contraceptive Research and Development Program (CONRAD) and the National Institute for Child Health and Development (NICHD).

The rather mechanical and molecular descriptions of the theory behind the action of immuno–contraceptives and the summary of the research carried out so far can make the whole process seem simple. The next two chapters provide an overview of the formidable hurdles that would have to be overcome if the immune system were to be employed for contraceptive purposes.

Notes

1. Spieler 1987:779.

2. For a good introduction to immunology, see Staines, Brostoff and James 1993. For challenging critiques of the molecular model of the immune system, see Haraway 1991 and Martin 1994.

3. I have used this example because it shows most clearly what may happen if the immune system reacts against the body's own tissues. However, the origin and development of rheumatoid arthritis is still being debated. Most immunologists believe that the auto-reactive immune cells are caused by an infection, i.e. that they are antibodies against particular bacteria which react against the lining because they 'see' it as similar to the original bacterial antigens.

4. I call this a 'menstrual-like period' rather than a 'period' so as not to obscure the difference between a natural period after nonfertilization of an egg cell and that after interception of hCG by antibodies.

2

Immuno-contraception: a feasible approach?

We are dealing with two of the most complex parts of the body, the immune system and the endocrine system. (N. Avrion Mitchison, immunologist, 1991)[1]

I am very skeptical that immunization against body constituents would ever work without side-effects. (David W. Hamilton, researcher at Minnesota University, 1990)[2]

David Griffin, manager of HRP's Task Force on Fertility Regulating Vaccines, has described how researchers came up with the idea of immuno-contraceptives:

> Following a detailed review of the options for the development of new methods, immunological intervention was considered a promising area for investigation, the objective being to use the body's own immune system to provide protection *in essentially the same way* that it provides protection against unwanted diseases. In other words to develop a fertility regulating vaccine. (Griffin 1993:37, emphasis added)

However, the development of immuno-contraceptives poses specific problems that were not encountered during the development of vaccines against unwanted infectious diseases, not least because of the fundamental differences between the desired immune responses and aims of immuno-contraceptives and those of anti-disease vaccines.

One of the few similarities between immuno-contraceptives and anti-disease vaccines is that their effects are both mediated by the immune system. In contrast to pharmacological contraceptives such as hormonal injectables and implants, the 'active principle [of immuno-contraceptives] is not what is injected but what is produced by the body in response' to the administered substance (Talwar et al. 1990b:585).

A crucial and profound difference between anti-disease vaccines and immuno-contraceptives is that vaccines advance our specific immune reactions against foreign micro-organisms, thereby increasing our resistance to specific diseases, whereas immuno-contraceptives induce abnormal immune responses against self or self-like body components or secretions essential for human reproduction. They induce an immune disorder: immune-mediated infertility.

There is a radical difference not only between the targets of the two

immune responses but also in the aims of the immunological technologies. The intended effect of anti-disease vaccination is the prevention
of harmful or life-threatening diseases. The stated purpose of immunological birth control methods, however, is effective and reversible
contraception over a predictable period of time.

Vaccines are effective not only because of their direct action in an
individual. Their efficacy also depends on the coverage of an immunization programme, that is, the proportion of immunized people in a given
area. If an immune response to a specific micro-organism is induced in
the majority of a vaccinated population, everyone gains from it, even
those people in whom the immune response may not be particularly
strong or who have not been vaccinated. Such persons are protected if
the majority of those around them are protected because the disease
will not be able to gain a foothold in the population and thus un-
vaccinated people are less likely to be exposed to the disease. But because
pregnancy is not a disease – particularly not a transmissible one – broad
coverage in a population does not increase the effectiveness at all for an
individual woman or man.

The duration of the immune response induced by a vaccine does not
need to be exact; the longer the immunological memory the better,
because the ultimate goal is lifetime protection against a disease.
Immuno-contraceptives, however, aim to induce an immune response
that is reliable in each and every individual and that occurs for a limited
period of time only. This is a complete novelty in terms of vaccine
technology. As Griffin states, 'unlike anti-disease vaccines, anti-fertility
vaccines are not intended to provide lifetime immunity but rather to
produce an effective immune response of predictable and comparatively
short duration' (Griffin 1990a:508).

Thus 'the longer the better' principle of anti-disease vaccines does
not apply to contraceptives, which people generally use when they want
to be temporarily infertile and may want to have children at a later date.
It should therefore be assumed that, in using a contraceptive, most
people would not want to risk lifelong, immune-mediated infertility.

The major undesired or unintended effects of anti-disease vaccines
are the outbreak of the disease itself in the immunized person and
immunological adverse effects such as the person developing allergies.
Potential undesired biological effects of immuno-contraceptives include
lifelong infertility and immune-mediated adverse effects to both user
and future offspring.

To develop an immunological birth control method, therefore, a
fundamentally new challenge has to be faced: the induction of a highly
effective, time-limited immune reaction to a specific reproductive component without inducing unacceptable adverse effects to user or future
offspring, a reaction which is reversible after a predictable period of time.

Antigens – safety and effectiveness

The major difficulty in designing an immuno-contraceptive is to find a target antigen whose neutralization by the immune system has no effect other than the desired, time-limited infertility.

A major concern is that the induction of an immune response to a human hormone or cell, such as is being attempted by the immuno-contraceptives under development, may cause an unintended immunological reaction against another body component which could lead to an *auto-immune disease* developing. To prevent such an immunological *cross-reaction*, researchers must ensure that the target antigen is immunologically dissimilar to any other body component. One way of attempting this has been to use only part (rather than the whole) of the reproductive component as the target antigen.[3]

If the target antigen is a hormone, it should not have any (significant) function in the body other than its reproductive function; if it does, the immune-mediated interference with the hormone's action may disturb not only its reproductive function but other body functions as well. To avoid the immune reaction damaging tissues surrounding the target component, such as egg cells or sperm cells, the location of the target antigen is also critical.

For the product to be effective as a contraceptive, moreover, the elimination of the target antigen must reliably interfere with fertility. The smaller the inducing antigen is, the less likely cross-reactions become – but it also becomes less likely that the method will prevent pregnancy. The major tension in 'vaccine' design has thus been to find an antigentic structure that does not induce immunological cross-reactions yet does induce sufficient antibodies to prevent pregnancy.

hCG (human chorionic gonadotropin) antigen   Many researchers consider the most promising target antigen for an immuno-contraceptive to be the pregnancy-associated hormone hCG (human chorionic gonadotropin). When current research into immuno-contraceptives began some twenty years ago, researchers believed that:

> the advantage of immunizing against ... human chorionic gonadotrophin (hCG) is that the antigen [hCG] is probably present in immunized women only at times of incipient pregnancy, and that risks of side-effects would therefore considerably be minimized. (Task Force 1978:368)

At that time, the hCG hormone seemed to be a 'perfect' antigen because it would need to be neutralized only once per month at most, that is, whenever an egg was fertilized. It was known that hCG was essential for the establishment and maintenance of early pregnancy and it seemed to have no function in a woman's body other than to prepare the uterus

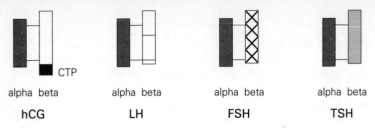

alpha beta   alpha beta   alpha beta   alpha beta

hCG   **LH**   **FSH**   **TSH**

hCG: Human chorionic gonadotropin
LH: Luteinizing hormone
FSH: Follicle-stimulating hormone
TSH: Thyroid-stimulating hormone
CTP: Carboxyterminal peptide

Figure 2.1 Similarities between hCG and three hormones of the pituitary gland (LH, FSH, TSH)

for an embryo's implantation. HRP, the Population Council and the Indian National Institute of Immunology (NII) all began research in the early 1970s into anti-hCG immuno-contraceptives.

Yet researchers were aware at this time that hCG does pose a problem in being a target antigen. Its molecular structure resembles that of other hormones: LH (luteinizing hormone), FSH (follicle-stimulating hormone) as well as TSH (thyroid-stimulating hormone). If the whole hCG molecule were used as an antigen, it would inevitably lead to immunological cross-reactions: the antibodies produced against hCG could also act against these other hormones because they are so similar to hCG. Such cross-reactions could interfere not only with the menstrual cycle but also with thyroid functioning and potentially lead in the long term to damage of the pituitary and thyroid glands.

To avoid this problem of cross-reactions, scientists broke down the hCG molecule to look for a more specific antigenic part. hCG and the related FSH, LH and TSH hormones are composed of two sub-units: a short alpha sub-unit and a longer beta sub-unit. While the alpha sub-unit is virtually identical in all four hormones, the beta sub-unit of hCG is similar only to that of LH. In addition, it has a small end section (called a carboxyterminal peptide or CTP which is composed of 37 amino acids) which, so far as researchers can determine, is not found on the other hormones.

HRP's anti-hCG immuno-contraceptive In order for HRP's immuno-contraceptive to act against the hCG hormone, the research group has developed a synthetic version of this CTP end section of the beta-hCG

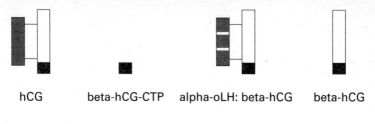

hCG   beta-hCG-CTP   alpha-oLH: beta-hCG   beta-hCG

beta-hCG-CTP	Carboxyterminal peptide portion of beta-hCG
beta-hCG	beta chain of the hCG hormone
alpha-oLH: beta-hCG	alpha chain of sheep (ovine) LH annealed to beta-hCG

Figure 2.2 hCG antigens used in current immuno–contraceptives

sub-unit to produce what it calls the beta–hCG–CTP– or 'peptide' vaccine.

Although this anti–beta–hCG–CTP formula does not appear to induce cross-reactions with LH in human trials (Jones *et al.* 1988:1,298), the preliminary animal (baboon) trial report mentioned unexpected cross-reactions between the blood serum of some of the baboons and laboratory tissue cultures of rat and baboon pituitary glands and pancreas (Rose *et al.* 1988:238–9). An article on the subsequent Phase I (safety) trial reported similar cross-reactions between some of the serum samples taken from women in the trial and laboratory cultures of baboon pancreas cells (Jones *et al.* 1988:1298 and HRP 1988:184). HRP Task Force manager David Griffin states that these adverse side effects were 'mere' artefacts of the test system' (Griffin 1994:2).[4]

The Population Council and National Institute of Immunology anti–hCG contraceptives   The Population Council and the NII have followed a different approach from HRP in developing their anti–hCG immuno-contraceptive. In their opinion, HRP's hCG fraction is too short ever to produce a sufficient immune response (Talwar 1994b:700).

The NII and the Population Council have therefore chosen to use the whole hCG beta sub-unit as the target antigen. This, as might be expected, induced immunological cross-reactions with LH in both animal and human trials. However, both teams claim that the cross-reactive auto-immunity induced did not result in anticipated adverse effects such as menstrual disturbances, and that cross-reactivity with LH is even beneficial because a concurrent action against LH may actually improve the immuno–contraceptive's efficacy.[5]

Immunologists from Johns Hopkins University and the University of Innsbruck queried the alleged lack of discernible effect of this cross-reaction at the 1989 meeting on immuno-contraceptives sponsored by HRP:

> Why isn't the bioactivity of hLH [human LH] impaired or even completely abolished in these subjects, when their immune responses are considered adequate to neutralize hCG? (Rose et al. 1991:131)

However, even using the whole beta sub-unit as the inducing antigen could not trigger a sufficient anti-fertility effect in early Phase I trials undertaken by the NII together with the Population Council during 1976–78. The immune response of one-quarter of the women did not reach the theoretically necessary level of antibodies to prevent conception (quoted in Jayaraman 1986:661).

After a number of other attempts to increase the efficacy in different ways, Pran Talwar, then director of the Indian NII, decided to change the target antigen still further. Instead of using the beta-sub-unit of hCG on its own, his team combined it with the alpha sub-unit of sheep LH (ovine LH or oLH). His rationale at the time was that the resulting combination antigen would resemble the whole hCG molecule more closely than just a part of the molecule would but that, unlike the alpha-hCG sub-unit, a sheep-based LH component would not lead to additional cross-reactions with the alpha sub-unit of human LH. The theory behind this claim was later disproved (Berger 1987; Ada and Griffin 1991:127). Despite knowing that the alpha-oLH:beta-hCG antigen is thus doubly cross-reacting with LH, the NII team continue to use it because they still claim it will be even more effective.

In countering concerns about the potential adverse effects of an anti-beta-hCG immuno-contraceptive cross-reacting with LH, the NII and the Population Council team refer to long-term animal studies carried out by the Population Council (Thau et al. 1987). However, according to HRP's research coordinator, David Griffin, it is far too early to make claims about the safety of immuno-contraceptives directed against the whole beta sub-unit because potential severe long-term effects, such as damage to the ovaries or pituitary gland, may occur or be detected only after years of repeated immunization. He emphasized that 'resolving this issue [of long-term adverse effects] through appropriately designed studies in relevant animal models should be a high priority in future research with this vaccine' (Griffin 1993:41).

The tension between the efficacy and the safety of anti-hCG products does not depend solely on the selection of the target antigen. The assumption that hCG is produced only by the early embryo and thus is a hormone specific only to the early phase of pregnancy has recently been undermined by the general finding that it may also be 'released in non pregnant women, presumably from the pituitary, in small amounts which vary with the menstrual cycle' (Mitchison 1990:725). When women's health advocates took up this question with researchers at the 1992 HRP Meeting on Fertility Regulating Vaccines, they were told:

> Research has established that the pituitary gland and certain types of lung cancers may also secrete hCG. It is not known if there are other elements in the body which also secrete hCG. The effect (beneficial or deleterious) that the immune response produced by anti-hCG vaccine may have on such hCG producing tissues is as yet unknown. (HRP 1993:17)

GnRH and FSH antigens Instead of hCG, several research teams are exploring the use of hormones not associated with pregnancy as target antigens, in particular, GnRH (gonadotropin-releasing hormone) and FSH (follicle-stimulating hormone).

GnRH and FSH act within complex feedback loops. GnRH's presence in the bloodstream determines whether the pituitary gland at the base of the brain releases the hormones FSH and LH, which in turn regulate whether a woman's ovaries release the hormones oestrogen and progesterone, and men's testes release testosterone. The presence of these hormones produced by the ovaries and testes in the bloodstream tells the brain – via the pituitary hormones FSH and LH – when to release more GnRH and how much.

All these hormones – GnRH, FSH and LH – are present continuously in our bodies. All have functions in the body besides the regulation of our reproduction. Thus it was known that choosing any one of them as the target antigen would increase the risks of the immuno-contraceptives having adverse effects.

Any interception of GnRH and FSH during their circulation in the bloodstream may interfere with all the biological effects of the hormones, not just with their effects on sperm or egg maturation. According to T. Chard and R. J. S. Howell, endocrinologists at St Bartholomew's Hospital Medical School in London, a 'very wide range of disorders' could potentially be caused by the hormonal imbalances induced (1991:97).

In the case of anti-GnRH immunization of men, the disturbance of testosterone production could cause a man to lose what are commonly identified as male characteristics, such as a low voice and body hair, and could also cause impotence – an inability to have an erection. In women, the neutralization of GnRH would affect the production of oestrogen and would cause a hormonal state that women usually experience only after menopause: in the short term, amenorrhoea and a potential decrease of libido; in the case of long-term GnRH neutralization, a potential increase in the likelihood of osteoporosis and circulatory diseases developing.

Some of the adverse effects, for instance atrophying of and damage to organs such as the ovaries or testes because of a lack of hormonal stimulation, might emerge only after long-term neutralization of GnRH. Researchers working on these types of immuno-contraceptives propose to offset the risks by concomitant administration of testosterone in men and oestrogen in women.

Any neutralization of FSH would not interfere with testosterone production in men, but the potential short-term adverse effects of interference with FSH's functions include weight increase, greasy skin, and breast development (gynaecomastia), whilst a potential long-term risk is prostate cancer (Chard and Howell 1991:97–109).

The real problem with anti-FSH immunization as a contraceptive, however, may be its lack of efficacy. It now seems that a man's body can produce sperm even in the absence of the hormone FSH (HRP 1990:30). Early on in this research, German reproductive researcher Eberhard Nieschlag advised against anti-FSH immunization for men because after a short period of low sperm production, his test animals gradually resumed sperm production (Nieschlag 1986).[6]

The adverse effects cited above may ensue simply from any interference with the biological function of hormones; they could, for instance, be caused by administering any substance that counteracts hormonal effects. But any of the above contraceptives, because they use the immune system, carry additional potential risks of affecting our immune system as well as our hormonal system and should, in fact, be thought of as *immune-mediated, hormonal* contraceptives.

The intended auto-immune reactions to prevent conception may act against these hormones not only while they are circulating in our blood, but also at two other stages: during hormone release and during the hormones' very attachment to the organs on which they are meant to act. GnRH and FSH may become bound by immune cells as they are being released from the hormone-producing cells in the brain, the hypothalamus and the pituitary glands respectively, or as they are attaching themselves to the pituitary gland or the testes respectively. While the hormone is in direct contact with a cell, the action of immune cells against the hormone might lead to cell destruction. This cytotoxic (cell-killing) effect 'might represent some of the most significant risks of immunization to self-antigens' (Chard and Howell 1991:99). Organ damage could occur if significant numbers of cells are affected by cytotoxic effects.

Thus auto-immunization against GnRH or FSH may not only interfere with a complex hormonal regulatory system but additionally carry the risk of direct tissue destruction. Experts at the 1989 HRP Symposium on Antifertility Vaccines concluded that 'the unknown consequences of chronic [long-term] immunity to "self" molecules in the brain, pituitary gland and gonads [ovaries/testes] would argue against using immunogens restricted to these sites' (Report 1991:255).

Egg and sperm antigens The development of safe immuno-contraceptives against sperm or egg cells poses different problems from those against hormones. The major difficulty is the need for the immune response to take place only in a very specific location.

Although in theory it seems elegant and safe to induce an immune reaction only once a month against just one, tiny body component, namely the (usually) one egg cell that ripens and is released from the ovary, researchers have in practice found it difficult to identify structures on the surface of the mature egg cells that are not also present on immature eggs within the ovaries. An auto-immune response against a structure that is present in both could eradicate all a woman's eggs, making her permanently infertile as well as damaging surrounding ovarian tissues. Such damage could adversely affect the production of oestrogen and progesterone by the ovaries.

Current research focuses on potential antigens on the egg's *zona pellucida* (ZP), a translucent membrane that surrounds the egg only in the late stages of maturation. However 'all of ZP vaccines tested so far in animals have produced reactions in the ovary ... that would make these current vaccines unacceptable for human use' (HRP 1993:41).

Immunization of men against sperm faces a similar problem. Immune reactions against sperm in the testes could cause a chronic auto-immune inflammation of the testes themselves (orchitis) which ultimately leads to male infertility. However, like eggs, sperm undergo changes over time. Researchers are therefore searching for post-testicular antigens: antigens that are expressed on the surface of the sperm cells shortly before ejaculation but after the sperm have left the testes, and expressed only then.

Many women who have taken the responsibility for contraception for years and borne the burden of various side effects may feel relieved that a potential anti-sperm contraceptive would free them from such effects. But to avoid problems of testicular inflammation, researchers are also developing an anti-sperm immuno-contraceptive that would act in women.

To act against sperm, a woman's immune system would have to be manipulated so as to produce a particular type of antibody which would have to be secreted into a woman's reproductive tract where the sperm would be. Each time a woman had sexual intercourse with a man ejaculating in her vagina, a large number of sperm would have to be neutralized. Some researchers are therefore proposing to identify sperm antigens that are present only when some of the sperm reach the Fallopian tubes – yet this can still be quite a significant number, necessitating the production of a huge number of antibodies. A final difficulty is that the desired immune reaction must be efficient, reliable and fast, no matter how frequently a woman has intercourse, since a single sperm cell can fertilize an egg within a few hours.

One problem with cell-associated antigens is similar to that with hormones: they could have similarities, some of them as yet unidentified, with other molecular structures in the body which could lead to unexpected cross-reactions following anti-egg or anti-sperm immunization. For example, some sperm antibodies have been found to cross-react with brain and kidney tissue, lymphocytes and red blood cells (Naz 1988).

So far, developing a safe and effective anti-egg or anti-sperm immuno-contraceptive has proved much more difficult than researchers anticipated, given the unusual and complex tasks that they are demanding of the human immune system. Although many potential cell-associated antigens have been screened for suitability, development of anti-egg and anti-sperm immuno-contraceptives is the least advanced of all the research into these novel contraceptive methods (Griffin 1993:41).

Immunization against body constituents: the underlying problem The challenge of finding a suitable antigenic component to interrupt a reproductive process should not be underestimated. Depending on the structure, function and location of the target antigen, different problems have occurred and will without doubt continue to arise when prototype formulas are tested in animals and humans. Some researchers claim that they will solve the tension between having an acceptably safe and yet sufficiently effective product by identifying more specific antigens. However, the basic principle of immuno-contraceptives would still remain: the induction of an immune response against a body substance that is usually protected by self-tolerance mechanisms. This will always have implications in terms of the immuno-contraceptives' projected effectiveness and safety.

Talwar and his colleagues at the NII expect that 'no single vaccine will be able to evoke a high enough response in all individuals' (Talwar et al. 1992a:1). Their suggestion is therefore to adopt a 'multivaccine strategy', that is, the 'delivery of multiple injections at a single contact point' because 'the utility of multidrug therapy in preference to monotherapy is recognized for chronic diseases such as leprosy and tuberculosis' (Talwar 1994b:702; Talwar n.d.:1). Others advocate the 'use of a combination of antigens from several reproductive molecules in a single "cocktail" vaccine' (Stevens 1992:139).

The NII team forgets to mention that 'multidrug therapies' run counter to the general trend in rational drug prescribing. Except for certain special cases, a prescriber should avoid giving more drugs than are absolutely necessary (nor should more than one ingredient be put into a single pharmaceutical product) in order to avoid a potential increase in adverse effects and interactions.

Both proposals – a multivaccine strategy and a single cocktail 'vaccine' – overlook the difference in target antigens, as well as in biomedical aims, between anti-disease vaccines and immuno-contraceptives. It is common practice to combine microbial antigens in anti-disease vaccines so that each antigen enhances the effects of the others. But while it may well be medically justifiable to create an anti-disease vaccine that acts concurrently against diphtheria, tetanus and pertussis (whooping cough) the case is entirely different for a contraceptive. The benefits of increasing contraceptive effectiveness must be measured against the increased risks from concurrent targeting of different reproductive components – and it should be remembered that other birth control methods are available.[7]

A fundamental safety question in terms of the design of an immuno-contraceptive will always be: *how will scientists know for sure that the antigen they select will not be seen by the immune system as similar to another body component?* How will they be able to exclude the possibility of potentially serious auto-immune diseases? The investigators who tested HRP's anti-hCG immuno-contraceptives in baboons found that 'each component of this complex vaccine mixture has the potential for inducing auto-antibody production and enhancing the level of pre-existing natural auto-antibodies'. The auto-antibodies in these short-term trials did not result in any auto-immune disorder. Yet, as the researchers stated, 'the fact that one cannot predict the target of these [auto-immune] reactions is worrisome' (Rose et al. 1988:231, 239).

Some scientists doubt that it will ever be possible to induce a sufficient anti-fertility effect without inducing unacceptable adverse effects. David Hamilton, a researcher in male contraception at the University of Minnesota, confronted his fellow researchers at the end of a 1989 CONRAD symposium on immuno-contraception by saying:

> We have heard during the meeting that zona pellucida antigens cause atrophy of the ovary … And in the male, even immunization with very sperm-specific surface antigens cause orchitis. Yet now it has been suggested … to get more specific antigens. But doesn't the inherent problem remain – that we are immunizing against body constituents and that this may cause auto-immunity? Although you may say we have examples already from human chorionic gonadotrophin (hCG) immunization, I think that these cases have not been followed properly. What do we know about those women who were immunized? Do you know what sort of delayed auto-immune disease is possible? I am very skeptical that immunization against body constituents would ever work without side-effects. (in Alexander et al. 1990:615)

Antigens – reversibility

Whether an increase in effectiveness will result in an increased risk is not the only challenge scientists face in developing immuno-contraceptives. To ensure the reversibility of immuno-contraceptives after a predictable period of time is another completely new aspect in designing a product that stimulates the immune system. In the case of anti-disease vaccines, the disease-preventing effect is prolonged not only when immunized persons receive a 'booster' injection of the vaccine but also when they come into contact with the disease-causing micro-organism.

For immuno-contraceptives to be reversible, however, it is crucial that the 'continuing or pulsed appearance of the target antigen ... should not act as a stimulus to the ongoing immune response' (Ada and Griffin 1991c:23). For example, in the case of an anti-sperm vaccine acting in women, while antibodies must be present in sufficient concentration to neutralize the sperm, the presence of the sperm themselves must not prolong the immune response; if they did, each time the immune system encountered sperm during intercourse, an immune reaction would be elicited, effectively making a woman permanently infertile.

For an immuno-contraceptive to be reversible, boosting based only on the original, *external* antigen-carrier construct should occur, so that if a woman or man wants to have a child and therefore wants to stop the effect of the contraceptive, she or he can decide not to have the next booster injection.

Clearly, one problem with immunization against body components, as opposed to foreign substances such as disease-causing micro-organisms, is that the natural body *internal* antigen may be present frequently, as is the case with the hCG hormone, or continuously, as is the case with non-pregnancy-associated reproductive hormones, and egg and sperm cells (although here the presence of the internal antigen depends on the ultimate candidate antigen: for example, the *zona pellucida* forms only during the monthly ripening of the egg cells). There is, therefore, a relatively high risk of continuous internal boosting, which could lead to lifelong infertility. Anti-sperm contraceptives in women are unpredictable because boosting would be dependent on coitus. (The theoretical risk of boosting is less with hCG, if hCG is present only after conception, that is, at most once a month.) Researchers are trying to avoid this problem by attempting to select an antigenic part of the reproductive component that would not stimulate the T-cells, the lymphocytes responsible for the immunological memory.

In the Phase II (efficacy) trial of NII's latest anti-hCG method, the contraceptive effect was, in fact, reversible; a number of women conceived after discontinuing use of the method (Talwar et al. 1994a: 8,533–4). However, whether all types of immuno-contraceptives will be reversible is not known, as immunologist Noel Rose stressed at the 1989 HRP symposium when he said 'boosting is the great unknown'.

A last factor which should not be underestimated is that the immune and endocrine (hormonal) systems are complex and not fully understood. How our bodies will react to deliberate anti-reproduction immunization in the long term is, as yet, unknown. For example, it is not known whether prolonged immunization against reproductive substances may lead to increased tolerance of these components rather than their neutralization (see, for instance, Rose et al. 1991:130). Immunologist Faye Schrater mentions this theoretical possibility in the case of anti-sperm immuno-contraceptives for women (Schrater 1994a:259).

Enhancing the immune response

> Vaccines to control human fertility in theory run a higher risk of inducing anti-self reactions compared with vaccines to control infectious diseases, and this risk may be greater still if powerful adjuvants are used and the vaccine is administered frequently. (Ada 1990:576)

In confronting the problem of how to reach a sufficient contraceptive effect without at the same time increasing the risk of adverse effects the choice of antigen is not the sole factor to be considered. Effectiveness as well as potential adverse effects also depend on the substances and procedures used to enhance the stimulation of the immune reaction, such as the carrier, the adjuvant, the 'vehicle' (water, oil or microsphere-based 'vaccine' formulation) and the immunization schedule.

To be effective, an anti-disease vaccine may need to be given more than once at carefully calculated intervals before a sufficient immune response against a specific micro-organism is generated. An immunization schedule usually consists of two parts: a primary schedule comprising the number of injections required to reach an effective level of protection; and a schedule for booster injections if a person wants to refresh the specific immune response once it begins to wane.

Immuno-contraceptives, however, should be given only as a single dose, that is, one injection or 'oral product' (for example, pill or liquid dose) which should create temporary infertility in all of the immunized women or men by inducing the primary immune response against the reproductive component for one to two years. If the user then wishes to continue the effects of an immuno-contraceptive, she or he would require a booster injection (or pill), that is, a renewed administration of the immuno-contraceptive to cause a secondary (tertiary, etc.) immune response for another one or two years.

The 'single administration' of an immuno-contraceptive is critical for several reasons. From multiple injections, there would be a potential risk of a person developing allergies (Basten et al. 1991:89; Griffin 1993:38) and, if the person did not receive all the primary injections, there would be a potential risk of foetal exposure, that is, there would be some immune response but not enough to prevent pregnancy.

It is known from experience with anti-disease immunization progammes that the higher the number of doses required for primary immunization, the fewer the number of people who will complete the primary immunization schedule. As Gordon Ada and David Griffin advised at the beginning of the 1989 HRP symposium, 'care is needed to make sure that any immunization schedule is completely effective as a contraceptive; in particular, an immunization schedule which is only partly successful may run the risk of fetal damage or incorrect development' (Ada and Griffin 1991c:22).[8]

None of the current prototype products has come close to this goal. Unsatisfactory efficacy is a major problem for all the anti-hCG contraceptives under development. HRP's prototype (which has a strong adjuvant and an oily 'slow release' formulation) requires two primary injections and booster injections every two to three months; the NII's formula (which uses a watery solution) needs three primary injections and boosters every three months (Stevens et al. 1990:563; Talwar et al. 1994a:8,533).

To rectify this problem, the HRP team and the NII researchers propose to incorporate their hCG–diphtheria toxoid conjugate and adjuvants into tiny biodegradable particles or microspheres which, once in the body, would result in a slower release of these immunologically active components than from the currently used oily emulsion. Whether or not this will prolong the action to one year – without at the same time decreasing effectiveness or increasing risks – is still unknown.

The number of primary immunizations and the ultimate duration of the contraceptive action are not the only uncertainties. USAID official Jeff Spieler warned that 'a fertility regulating vaccine ... would have to produce and sustain effective immunity in at least 95% of the vaccinated population, a level of protection rarely achieved even with the most successful viral and bacterial vaccines' (Spieler 1987:779).

The percentage of women who will never have an effective immune response is likely to be relatively high because immune reactions depend on our genetic predisposition. In disease prevention, wide genetic differences have helped members of the human species to adapt to their different environments to the optimum degree.

As already mentioned, in the case of anti-disease vaccines, lower effectiveness in terms of antibody production can be offset by increasing the number of primary immunizations and by increasing the coverage of the immunization programme so as to lower the potential spread of a particular disease. Both of these measures cannot apply to immuno-contraceptives.

The Population Council, NII and HRP have all pursued different strategies to solve the tension between efficacy and safety. Each of them has chosen a different antigen: the Population Council the whole beta-hCG unit; NII the whole beta-hCG unit coupled to a sheep alpha-LH unit; and HRP a fraction of the beta-hCG. The Population Council coupled the antigen to a tetanus toxoid, added alum, which is a long-established adjuvant in anti-disease vaccines, and water – and did not achieve a satisfactory immune response. In its latest Phase I (safety) trial of its anti-beta-hCG prototype, a 'great variability in [immune] response among subjects was noted. A significant proportion of subjects had a low and short lived response.' The team hoped to 'improve and prolong the response' through the 'addition of an adjuvant [additional to alum]' (Brache et al. 1992:9–10).

The results of the NII Phase II (efficacy) trials cast doubt upon the possibility of improving the Population Council's formula simply by adding one more adjuvant. The NII had started, as the Population Council did, with alum in water. But in its Phase I (safety) trial with different hCG antigens, 15 per cent of the women in the trial did not react to the primary refresher injection. This non-responsiveness was attributed to the fact that the women had recently been immunized against tetanus and had thus become tolerant to the tetanus carrier (Gaur et al. 1990). To circumvent this, the immunization schedule in the Phase II (efficacy) trial with the alpha-oLH:beta-hCG antigen became incredibly complex: an additional adjuvant was added to the first of the three primary immunizations while the carrier changed from tetanus to diphtheria and back to tetanus, a switch described by Talwar as an 'alternate carrier' strategy.

Given the low efficacy of even the whole beta-hCG unit, the NII decided to anneal part of the ovine luteinizing hormone (LH) to its antigen, thereby increasing the risk of cross-reactions. Yet, in the Phase II trials of this anti-hCG-oLH immuno-contraceptive, 20 per cent of the trial participants never reached the threshold antibody level required to prevent pregnancy (Talwar et al. 1994a:8,535).

HRP, on the other hand, in order to offset the anticipated low immune response to its small hCG fraction, added a relatively strong adjuvant, muramyl dipeptide (MDP), which is not approved even for anti-disease vaccines.[9] Moreover, it formulated the product as an oily emulsion, not a watery solution, in order to increase the duration of immune stimulation by slowing down the release of the hCG-CTP diphtheria toxoid conjugate from the injection site, in this case women's buttocks.

However, there were problems in both Phase I (safety) and Phase II (efficacy) trials with this formulation. In the Phase I trials, several women experienced transient muscle and joint pains within forty-eight hours of the injection. At the time, the team concluded that the oily emulsion must have been unstable,[10] that is, the adjuvant must have been released too fast and thus in too high a concentration. The ten women with the greatest joint and muscle pain were replaced in the trial by women receiving a supposedly more stable product (Jones et al. 1988:1,297–8).

Despite its attempt to reformulate the product for the Phase II trial, the problems were not solved. In June 1994, HRP suspended the Phase II trial in Sweden 'following the occurrence of unexpected but transient side effects in the majority of the first seven women volunteers admitted to the trial' (HRP 1994b:8). The principal investigator, Marc Bygdeman from the Department of Obstetrics and Gynaecology at the Karolinska Hospital in Stockholm, made this statement:

> All seven women had moderate to strong pain locally at the injection site [i.e. the buttocks], sometimes radiating down the leg for 1–3 days. The problem could not be solved by reducing the dose given or by administering the dose at two injection sites. Four of the patients had a slight temperature elevation which lasted for a few days. Two women developed a sterile abscess at the injection site. (Bygdeman 1995)

Immuno-contraceptive feasibility

Can the tension between contraceptive efficacy and safety ever be resolved? Some researchers refer to naturally occurring immunological infertility or to the fact that many people have antibodies against their own body components without any apparent disease or adverse symptoms. As John Aitken at the Reproductive Biology Unit at Edinburgh University says:

> We have a rather exciting clinical model: it's the five per cent of men whose infertility is caused by a spontaneous immunity to their own sperm while some women become immune to their partner's sperm and neither suffer any side effects. Clinically this immunity is a problem, but nature is doing exactly what we want to be able to replicate and reverse. (Hope 1992:29)

However, infertility caused by a woman's spontaneous immune reaction against sperm, for example, is profoundly different from an artificially induced one. The 'natural' immune response against the reproductive component does not need to be enhanced.

Caution over the very concept of immunological contraception seems appropriate in view of the fact that, although studies over the past twenty years have showed that immunization with reproductive components alone can cause some degree of infertility, they have also indicated that as soon as these molecules were formulated as prototype 'vaccines', their increased effectiveness was invariably accompanied by various types of disturbances and/or an increased risk of immunological adverse effects (Griffin 1990a:503).

Notes

1. Mitchison 1991:247
2. In Alexander et al. 1990:615.
3. A reproductive component has many different antigenic determinants. The more antigen determinants there are in a reproductive component, the more different types of antibodies are likely to be generated by an immune response induced by an immuno-contraceptive that targets the whole component – and the more cross-reactions there are likely to be. Choosing just part of the reproductive component as the inducing antigen may reduce the likelihood of cross-reactions – but also reduces the potential effectiveness of the method in preventing pregnancy.
4. The published report of the 1992 HRP Meeting on Fertility Regulating Vaccines specifies that 'there was no consistent pattern of these reactions, in that they were observed with both pre-immunization and post-immunization samples, and that there was no correlation between the results obtained by two different laboratories sent the same serum samples' (HRP 1993:17).
5. Some researchers involved in the last Population Council trial stated that 'if any, the cross-reaction with LH will cause an additional anti-fertility effect by impairment of luteal function' (Brache et al. 1992:10).
6. Reviewers at the 1989 HRP Symposium on Assessing the Safety and Efficacy of Vaccines to Regulate Fertility stated: 'However, spermatogenesis returned over a period of 4 years. With the realization that spermatogenesis can be maintained with testosterone alone, selective FSH suppression is unlikely to be a viable contraceptive for use by men' (Chard and Howell 1991:103).
7. The expert speaking on the 'regulatory aspects of vaccine development' at the 1989 Symposium on Fertility Regulating Vaccines stressed the biomedical differences between anti-disease immunization and immuno-contraception. 'It should ... be anticipated that much stricter requirements will apply to an anti-fertility vaccine, in view of the target antigen [being a human body component], the fact that it may be administered more frequently than a conventional anti-microbial [i.e. anti-disease] vaccine, and because the vaccinee population consists of women of child-bearing age' (Weber 1991:194). This assessment, however, does not consider that other contraceptives are already available, a fact that has to be taken into account when assessing the risks and benefits of immuno-contraceptives.
8. Ada and Griffin pointed out that this concern may be most relevant in the case of immuno-contraceptives directed to a product or products of the gametes or the conceptus, i.e. sperm and egg cells, and pregnancy-related hormones such as hCG.
9. Despite research to find safe adjuvants that would elicit a better immune response, alum is still the only adjuvant approved by the US FDA for use in human vaccination.

10. This is similar to an everyday observation of salad dressings in which vinegar is the watery ingredient to be emulsified in oil. If the emulsion breaks up, the vinegar and oil may separate. In the case of 'slow release', oily immuno-contraceptive emulsions, the immunologically active 'ingredients' are in the watery solution. If the emulsion breaks up, they will be released much quicker and thus cause a different immune reaction.

3
Biomedical characteristics and risks

Antifertility vaccines ... are given to healthy people to prevent a normal, non-disease-inducing physiological process. If pregnancy does occur after vaccination, the ongoing immune response may constitute a risk to the developing fetus. Furthermore, there is a variety of alternative methodologies for the same purpose. (Gordon Ada and David Griffin, writing about the evaluation of immuno-contraceptives, 1991)[1]

The immune system does not operate in isolation. This is one of its cardinal features ... In order to understand the immune system it is important to understand that it interacts in a dynamic way with all the tissues of the body. As a consequence of these interactions, the proper functioning of the immune system is essential to the proper functioning of the body as a whole. When the immune system malfunctions, other parts of the body may suffer, and in turn, disturbances elsewhere may perturb the immune system and influence our resistance to infection. (Immunologists Norman Staines, Jonathan Brostoff and Keith James, 1993)[2]

Many women wish to have access to a range of contraceptive options according to their different preferences at different stages in their lives.[3] These will depend, among other factors, on their circumstances, freedom to make their own decisions concerning their bodies and their lives, economic independence, the number of children they wish to have, their age and health, and access to health facilities including safe abortion services. A variety of modern contraceptive methods are now available, while various practices such as withdrawal and breastfeeding can also have a contraceptive effect. Could immunological birth control methods offer users anything that existing contraceptives cannot and thereby positively enhance contraceptive options?

A prospective risk–benefit assessment

Many researchers tend to state that it is too early to raise this question at the current state of research. They maintain that the immuno-contraceptives used in clinical trials are only early prototypes and that the final products are likely to differ substantially from what has been tested so far (see, for instance, Griffin, Jones and Stevens 1994).

In my understanding, however, of medical research ethics, it is never too early to assess the probable benefits of a medical product or procedure and to weigh them against the potential risks, while bearing in mind other methods and practices that achieve the same end. An assessment of the likely risks and benefits should accompany the development of any biomedical product right from the very germ of the idea through to the decision to go ahead with each trial phase and through to the product's approval for the market by national regulatory drug authorities and after. Assessing the advantages of a new contraceptive under development over existing methods goes well beyond simply evaluating the risks for participants in each particular trial. It means projecting the birth control method into the future and considering the real life conditions of the majority of the future users – or 'targets'.

Any evaluation of the advantages of this new class of immuno-contraceptives over existing methods should include an attempt to outline as far as possible its future likely biomedical characteristics and then to apply criteria to assess the anticipated benefits and anticipated risks of these products.

The choice of criteria in such an assessment is as critical as considerations of the likely characteristics of a contraceptive and of external factors not inherent in the technology. It is the choice of criteria that determines which of the anticipated consequences of the development of the technology are considered as risks and which as benefits.

This question of criteria has preoccupied me intensely since 1988, when I started working on various health aspects of contraceptives as a team member of a Thai consumer protection group. Women's health activists and feminist researchers have long pointed out that criteria of what constitutes a 'good' contraceptive are as various and as contradictory as different societies' views on women's rights to self-determination over their reproduction and sexuality (Morgall 1993; Wajcman 1994; Hartmann 1995). In a user-centred analysis, the following criteria seemed most appropriate both to me and to those who reviewed this publication.[4]

In terms of its biological effects, a contraceptive should prevent pregnancy reliably; the effects should be reversible; and neither short- or long-term adverse effects should pose major risks to the health of the person using the contraceptive. Because contraceptives are used by large numbers of healthy, not sick, people over long periods of time, safety standards should be more stringent than those employed for drugs intended to treat diseases, or for vaccines to prevent them. In addition, there should be no risk to the health of any offspring born after contraceptive use, nor to any baby born during contraceptive use if the method fails.

At a minimum, from the perspective of someone using the contraceptive, a medical risk–benefit assessment of a contraceptive should consider the following:

— the method's effectiveness and reliability in preventing pregnancy;
— the risk of causing permanent infertility;
— the likelihood of short- and long-term adverse effects, including rare but serious adverse effects. For instance, hormonal contraceptives can cause potentially fatal heart attacks and strokes and, more frequently, 'minor' side effects such as headaches or menstrual disturbances. Not only the safety of a birth control method but also its general impact on a user's well-being should be considered;
— the potential effects of the contraceptive on a foetus in cases where the method fails to prevent pregnancy;
— the method's effect, if any, on the transmission of sexually transmitted diseases including HIV;
— the health service requirements for safe use of the method.

However, the likely risks and benefits of a contraceptive cannot be measured solely in biomedical terms. An evaluation of the advantages of a contraceptive over other birth control methods should also include important nonmedical considerations, in particular:

— the method's effects on social and sexual relationships;
— the potential for the method to be abused, that is, to be used coercively or without the fully informed consent of the person receiving it.

These aspects are considered in Chapter 4.

In making a biomedical risk–benefit assessment of immuno-contraceptives, their probable characteristics can be extrapolated from knowledge obtained from the animal and human trials of the various immuno-contraceptive formulations, and from what is known about both the immune system and its characteristics, and the various responses of the body to anti-disease vaccination.[5]

The balance between probable benefits and risks, however, has to be assessed against that of other contraceptives, never against anti-disease vaccines. This is because the criteria appropriate to assess a medical technology that may prevent a harmful disease are profoundly different from those appropriate to assess a contraceptive technology that may prevent a healthy process and to which there are alternatives.

Immuno-contraceptives rely on our immune system, a fact that has important and distinct implications in terms of their potential benefits and risks.

Contraceptive effectiveness

No contraceptive is 100 per cent effective in preventing pregnancy – although some come close. Effectiveness is generally measured as the

Table 3.1 Comparison of some methods of contraception

Method	Pregnancy rate per 100 women years[a]	Risk of adverse effects	Protection against STD/HIV?
Sterilization			
Male	0–0.2	Low	No
Female	0–0.5	Medium	No
Oral hormonal contraceptives			
Combined pill	0.2–7	Medium	No
Progestogen only	0.3–5	Medium	No
Hormonal injectables	<1[b]	Medium–high[e]	No[f]
Hormonal implants	0.3–1.4[c]	Medium–high[e]	No[f]
IUDs	0.3–9	High	No[g]
Diaphragm	2–20	Low	Yes/No[h]
Condom	2–20	Low	Yes
Sponge	9–27	Low	Yes/No[h]
Spermicide	4–30	Low	Yes/No[h]
Withdrawal	5–20	–	No
Rhythm	25–30	–	No
Ovulation monitoring	3–25	–	No
Breastfeeding	2[d]	–	No

Notes: [a] Large variations in pregnancy rates for some methods mostly reflect differences in how consistently or how well a method is used. [b] The low pregnancy rate with injectables is dependent upon consistent use; this is more likely to occur in clinical trials than in real-life situations. [c] The pregnancy rate with implants is higher than this in women weighing more than 70 kilograms. [d] Breastfeeding makes a substantial contribution to birth spacing and fertility control in many areas, and gives a pregnancy rate of less than 2 per 100 women years during the first six months provided the baby is nearly fully breastfed, and the mother's menstruation has not yet returned. [e] The long-term safety of injectables and implants is not established; also any adverse effects may continue for the duration of the effect of the injection or until the implant has been removed. [f] Injectables and implants may increase the risk of transmission of HIV if unsterile equipment is used. [g] The use of IUDs is probably a risk factor in some STDs. [h] The diaphragm and some other barrier methods reduce the risk of transmission of some STDs, but have no effect in preventing the transmission of HIV. *Source*: Chetley 1995:246

number of pregnancies occurring in 100 women who use a particular method for one 'woman' year (12 menstrual cycles of 28 days). *Theoretical rates of effectiveness* are based on the results of controlled clinical trials, carried out under ideal conditions. *Actual rates of effectiveness* reflect conditions of actual use of a method.

The actual or reported effectiveness of contraceptives varies according to different studies (see Table 3.1), a variation that may reflect, among other factors, the underlying fertility rate in the group of people participating in the studies. This underlying fertility is related to age, health and frequency of sexual intercourse and to other factors that influence the effectiveness of a particular contraceptive method. For instance, Norplant® is less effective if a woman weighs more than 70 kilograms.

HRP's David Griffin told women's health advocates at the 1992 HRP meeting that 'fertility regulating vaccines would need to be as effective in preventing an unwanted pregnancy as the best of the currently available alternative methods' (Griffin 1993:38). But because of genetic differences between people, it is unlikely that a particular immuno-contraceptive will be effective for more than 95 per cent of users (Spieler 1987).

As Table 3.1 shows, a method failure rate of 5 per cent would be *higher* than the minimum failure rates recorded for those contraceptives often rated by family planners as relatively ineffective, such as condoms, diaphragms and some 'natural' methods including ovulation monitoring and breastfeeding on demand.

Efficacy of immuno-contraceptives

'Model' versus 'actual' efficacy   How will a user of an immuno-contraceptive know whether the method is acting sufficiently to prevent pregnancy? As the action depends on antibodies to the target antigen, the only way to determine whether the immune response to reproductive components is sufficient is to measure the amount of antibodies to the target antigen in the blood, known as the *antibody titre*. For an immune response to be sufficient to prevent pregnancy, this amount needs to be at or above a certain threshold level.[6]

Figure 3.1 shows the theoretical progression of effectiveness of a 'model' immuno-contraceptive, given as a single injection or oral dose, and reversible after a period of one to two years:

— After injection (A), the body begins to build up its immune response to the target antigen. The period of time from injection until a sufficient immune response has been generated to reach the antifertility threshold (B) is called the *lag period*. During this time, the immune response cannot be relied upon to prevent pregnancy.

— When the amount of antibodies reaches the antifertility threshold (B), the immune response is sufficient to prevent pregnancy until it slowly decreases to below the threshold (C). The period when the immune response is at or above the threshold is the *contraceptive phase*.

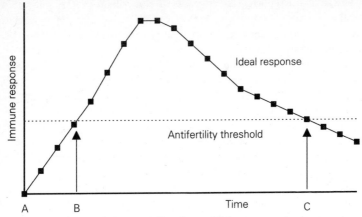

Figure 3.1 Efficacy profile of a model immuno-contraceptive

— After the antibody titre has dropped below the effective threshold level (C), the contraceptive effect is no longer reliable and the immune response enters the *waning phase*. There is no antifertility effect, unless the immune response is prolonged through a booster injection of the immuno-contraceptive.

The efficacy of an immuno-contraceptive in preventing pregnancy thus depends on the lag phase, the duration and reliability of the contraceptive phase, the waning phase and boosting.

This description of the efficacy profile of a model immuno-contraceptive, however, obscures the high frequency of deviations to be expected in actual use by different women in different circumstances:

> In common with other [i.e. anti-disease] vaccines, the immune response elicited by fertility regulating vaccines will vary from one individual to another and will depend on the constitution and the genetic, nutritional and health status of the user. (HRP 1993:18)

It is known from studies of anti-disease immunization that the speed and progress of the immune response varies between individuals. Figure 3.2 shows the variety of y immune responses that could be expected from any immuno-contraceptive. Not all of those receiving the contraceptive would reach the threshold at the same time, that is, the duration of the lag period would vary. Some people would not reach the antifertility threshold at all (●) while others would barely reach it or reach it for a very short period of time only (▲). For this group of persons, any disturbance of the immune system, for instance as a result of illness, malnutrition or stress, could lead to a drop in the antibody titre below the threshold.

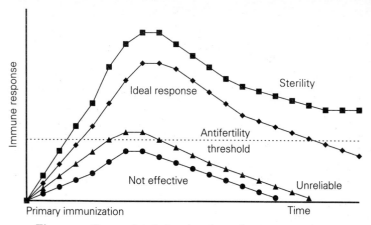

Figure 3.2 Expected variations in primary immune responses to reproductive components

The duration of the contraceptive phase could also differ; the immune response of some people (for example, people with a predisposition to allergies or auto-immune disease) may not even go into the waning phase, meaning that they would be sterile for the rest of their lives (■).

Lag period There is a lag period for all immune responses, including those triggered by micro-organisms, anti-disease vaccines and immuno-contraceptives, which cannot be avoided because of the nature of the body's immune responses. During this time, users would have received and been exposed to the immuno-contraceptive but it would not yet be effective as a contraceptive.

But, as Figure 3.2 shows, the duration of the lag period would vary in different women. It would also depend on the type of immuno-contraceptive. For example, although HRP and the NII are both using hCG as an antigen, the lag periods with their different preparations are different.

In the case of HRP's current formula targeted against hCG-CTP, an antibody level considered effective against pregnancy takes around five to six weeks to build up (Jones et al. 1988). Talwar and his colleagues at the NII state that their anti-hCG immuno-contraceptive

> takes about 3–4 months to build up antibody levels above the protective threshold. This period will be vulnerable to pregnancy and it is important to devise an approach which is compatible with the ... vaccine for covering this lag-period. (Talwar et al. 1992a:5)

To cover the lag period, Talwar's team is developing a contraceptive based on purified extracts of the neem tree to inject into a woman's uterus at the same time as the first administration of the immuno-contraceptive. Neem seed extracts have been used as pesticides and vaginal contraceptives in India for centuries. The contraceptive effect after intrauterine administration probably results from inflammation of the uterus. The team claims that this substance causes a highly localized contraceptive immune reaction for a few months. Called Pran (Talwar's) Neem-based Vaccine for Inducing Local Cell-mediated Immunity or Praneem VILCI, it is being promoted by the team not only as a 'companion vaccine' to its anti-hCG contraceptive but also as a 'vacation contraceptive' (Talwar et al. 1993:6). Despite its name, a herbal extract injected into a woman's uterus is neither an immuno-contraceptive nor a vaccine.

The NII team tested this neem extract in eighteen Indian women and then planned to test its anti-hCG formula combined with Praneem VILCI to act as a lag-period contraceptive (Talwar 1994a:2). Such a trial, however, is of dubious scientific quality because it would not be possible to distinguish which contraceptive effects or adverse effects were caused by the immuno-contraceptive and which by the intrauterine neem extract.

In addition, it should be questioned whether generalized uterine inflammation is acceptable as a means of contraception and whether, in real life conditions, such intrauterine insertions would be carried out in the careful and sterile manner needed to avoid the risks of pelvic infection.[7]

The NII's neem extract notwithstanding, the problem of contraception during the lag period of immuno-contraceptives remains. If a woman who had received an immuno-contraceptive became pregnant during the lag period, the foetus would be exposed to the immune response induced by the contraceptive because ongoing immune responses cannot just be 'switched off'. The consequences of such exposure for the development of the foetus, if the woman decides to continue the pregnancy or does not have access to an abortion, are as yet unknown. They may become known only after thousands of women have received immuno-contraceptives.

HRP's task force manager David Griffin has stated that:

> this [lag period] is not perceived as a problem that would make the vaccine unattractive. Synchronizing vaccine administration with the menstrual cycle and/or using the monthly injectable contraceptive for the first month are two of several possible strategies that have been considered for this context. (Griffin 1990b:6)

Synchronizing the administration of an immuno-contraceptive with women's menstrual cycles makes sense, however, only if the lag period of immuno-contraceptives could be decreased to less than ten days – the next earliest possibility of conception after the last menses (depending on an individual woman's cycle). This seems a rather remote possibility for immuno-contraceptives. Indeed, Vernon Stevens from Ohio State University and the originator of HRP's anti-hCG contraceptive makes the following prediction:

> It is well known and widely accepted that significant antibody production following successful immunization does not occur until 7–10 days following exposure of the lymphoid system to an immunogen. In practice, responses to weak antigens [such as reproductive components] usually require 14–21 days. (Stevens 1992:139)

Even if the lag period could be reduced to less than ten days, care would still have to be taken that the immuno-contraceptive was indeed given at the beginning of the menses – rather than just on any day when a woman came to the clinic or hospital.

Because any ongoing immune response is a potential risk to a foetus, any contraceptive used in the lag period has to be highly reliable and effective – but should not interact with the action of the immuno-contraceptive. As it is not known whether and to what degree hormonal contraceptives decrease the immune response, a woman or man may need to rely on barrier methods in the lag period. Vernon Stevens has postponed a definitive statement on the birth control methods to be used during the lag phase:

> At the point in time when immunological birth control methods are ready to enter family planning programmes, this issue will need to be seriously addressed. (Stevens 1992:139)

Contraceptive phase

> The development of new [antifertility] vaccines is a particularly difficult exercise in view of the *unpredictability* of individual immune responses, and the fact that effective immunity depends on an active and ongoing process of long duration. (Griffin and Jones 1991:189; emphasis added)

Like the lag period, the duration and magnitude of the contraceptive immune response will also vary, depending on a person's immune system and circumstances.

Trial reports to date do not give a clear indication of the extent of the variations. Both HRP's Phase I (safety) trial and NII's Phase II (efficacy) trials give the duration only in terms of mean value.[8] For example, the HRP trial (which divided trial participants into five groups of four women, each group receiving differing doses of anti-hCG antigen and adjuvant) reports:

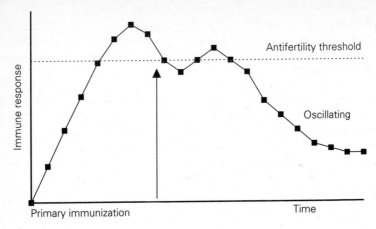

Figure 3.3 Drops in contraceptive effect due to illness or stress

The mean values [of anti-hCG antibodies] in groups 1–4 ... persisted for almost six months. In group 5, the mean value ... remained above [the contraceptive threshold] after six months. Longer term follow-up of two subjects in group 5 has indicated the persistence of a contraceptive antibody level for 9 and 10 months. (Jones et al. 1988:1,297)

NII researchers report that in its Phase II trial, 'booster injections ... were given at an average of 3 months' (Talwar et al. 1994a:8, 533).[9] The variations in duration of the contraceptive phase thus cannot be identified from the reports of either the HRP or NII trials – but both reports indicate that the respective durations are as yet certainly much shorter than the intended one to two years.

Because the duration and magnitude of the imune responses depend on the immune system, a person may have a low immune response because his or her immune responses are suppressed by malnutrition, stress or severe infections such as malaria, tuberculosis, hepatitis or HIV. In addition, certain pharmaceutical medicines, such as corticosteroids and anti-cancer drugs, are immunosuppressive (Basten et al. 1991:78).

It is difficult to predict the extent of these types of low immune response. At any time while using immuno-contraceptives, a woman can be exposed to a disease or greater life stresses. The problem of low responses caused by external factors is that their onset will be unpredictable. When disease or life stress occurs, the antibody titre may drop below the effective level without the user being aware of this, as illustrated in Figure 3.3. The only way to know would be to test the antibody level.

Waning phase and boosters After a certain (unpredictable) length of time, the effect of an immuno-contraceptive should wear off as the level of antibodies drops. This is the same process as occurs with some anti-disease vaccines, such as tetanus, which require booster injections to maintain the protective antibody titre.

So that women would know when their antibody level was insufficient and that they should use additional contraception or seek a booster injection of the immuno-contraceptive, some sort of test would be needed. (There is also a lag phase after a booster injection, although it is shorter than that for primary immunization.) HRP's David Griffin believes that testing would not be a problem:

> Individual differences in the length of period of effective immunity are to be expected with these preparations [antifertility vaccines]. This will require the development of simple, reliable and inexpensive means of monitoring the immune status of vaccine recipients. However, this is not seen as a problem as several reliable and inexpensive clinic-use (and maybe even self use) antibody measuring kits, requiring only a skin-prick blood sample, are under development. (Griffin 1990b)

Alternatively, Griffin suggests that this problem of a period of insufficient and indeterminate immune response could be overcome by identifying the minimum effective period for an immuno-contraceptive and advising women to have a booster or use an alternative method before they reach a potential waning phase (Griffin 1993:42). However, if this mimimum effective period is short, entailing frequent booster injections, several potential risks may be increased, including that of allergic reactions and permanent infertility.

Reversibility The ultimate duration of the anti-fertility effects of an immuno-contraceptive would depend not only on the type of immuno-contraceptive. Even if internal boosting from the immune system encountering the natural antigen could be excluded safely, it does not follow that the anti-fertility effect would be reversible. Two more factors have to be considered.

First, reversibility may also depend on the duration of use. It is as yet unknown whether and to what degree prolonged, repeated use of immuno-contraceptives may ultimately lead to irreversible immunological infertility.

Second, certain groups of persons may be at special risk of permanent infertility from immuno-contraceptives, in particular persons with a predisposition towards allergic or auto-immune diseases. They may experience 'an excessive response [to an immuno-contraceptive] ... resulting in irreversible infertility' (Basten et al. 1991:78).

The extent of this risk will only be known after Phase III large-scale trials which aim to assess the efficacy and risks of contraceptives in the general population in normal family planning service conditions. The problem in excluding such risk groups, however, is that a predisposition to these immunological disorders cannot always be predicted from past medical or family history (Basten et al. 1991:79).

Reversal on request   If someone using an immuno-contraceptive wished to have a child before the contraceptive effects wore off, little could be done at present. As Vernon Stevens concedes, 'our current state of knowledge offers no safe means of terminating the effects of immunization upon demand and returning their fertility immediately' (Stevens 1992:142). HRP's David Griffin, in responding to a colleague's comment about the lack of priority accorded to the need for immediate reversibility in the research and development of immuno-contraceptives, had this to say:

> The question [of] reversal on demand ... would be dealt with in the same way that you would, for instance, deal with an injectable steroid. You would inform the recipients that they would have to accept that the vaccine would be active for the stated duration of time, whether one year or two years, or however long. And you would also inform them that it is irreversible during that period. (Griffin 1990a:521).

Griffin proposed, however, to attempt to counteract the antifertility effects of anti-hCG contraceptives by administering progesterone (Griffin 1993:42). This would not interrupt the immune neutralization of hCG; it would simply replace the progesterone secreted by the ovaries with a synthetic or natural progesterone product, so that if a woman became pregnant while the anti-hCG contraceptive was still effective, the administration of progesterone might lead to a continuation of pregnancy rather than the shedding of the uterine lining.

However, this countering measure would not exclude potential adverse effects for the woman or her foetus. Exposure to synthetic progestins is not recommended during pregnancy because of the risk of malformations (Australian Drug Evaluation Committee 1989), whilst 'natural' progesterone products are frequently assumed to be safer, an assumption based on very little evidence.

David Griffin's more recent proposal is to administer 'free' unbound hCG-CTP (that is, the antigen but not formulated as an immuno-contraceptive) in excess so as to 'swamp' anti-hCG antibodies. For both these proposals, however, Griffin cautioned in 1993 that 'at present it is not known if these procedures could be advocated for clinical use' (Griffin 1993:42).[10]

Both these proposals – administering progesterone or hCG-CTP – are applicable only to the reversal of anti-hCG immuno-contraceptives.

For each different immuno-contraceptive type, another method would have to be found and tested.

The desire to have a child is not the only reason a woman or man may wish to stop using a contraceptive. What if they experience severe side effects, such as serious auto-immune disorders? In such cases, it would not be possible to switch off the immune response. People would have to wait until the effect wore off or use drugs to suppress their immune system.

This lack of immediate reversibility is a serious drawback of immuno-contraceptives when considering the real life situations of potential users. In responding to criticisms of immuno-contraceptives, Griffin has stated that if serious auto-immune disorders were discovered during trials, HRP would discontiue research into the contraceptives (Griffin 1990b). The duration of action and reversibility on demand are not only safety concerns for the user and any potential foetus; they also affect the abuse potential of immuno-contraceptives (see Chapter 4).

Adverse effects

> Anti-fertility vaccines are not intended to combat life-threatening or de-bilitating diseases, but are to be used by healthy, fertile individuals who have a number of alternative existing family planning methods from which to choose. These considerations necessitate that the assessment of safety is of paramount importance in the development of these vaccines. (Ada and Griffin 1991a:xv)

As with the efficacy profile of an immuno-contraceptive, the fact that this method relies on the working of the immune system and, moreover, that its effect is based on immunization against 'self' or 'self-like' human components has implications in terms of the range of potential adverse effects. These include auto-immune diseases, allergies, immune-complex diseases, and interaction with and exacerbation of existing diseases.

Auto-immune diseases As some researchers have pointed out, the presence of antibodies against body components – auto-antibodies - does not necessarily mean that an auto-immune disease will develop (Ada and Griffin, 1991b:6–8). David Griffin and Warren Jones state that, 'It is not clear whether auto-immunity, either naturally occurring or vaccine elicited, is predictive of future auto-immune disease' (1991:186). The team testing the current HRP product in baboons concluded: 'We do not feel ... that production of auto-antibodies by itself is a contra-indication to *careful* testing of an hCG subunit vaccine'. But they also stressed, 'The possibility that prolonged immunization in genetically predisposed individuals might lead to immunopathic [disease-causing] complications must remain a consideration' (Rose et al. 1988:231, 239; emphasis added).

In practice, it would be very difficult to define and to identify 'genetically predisposed individuals' who should not receive an immuno-contraceptive because they might develop an auto-immune disorder. Auto-immune diseases tend to be more frequent and more severe in women than in men (Playfair 1989:33).

Although immunologists are working hard to find cures for auto-immune diseases, it is questionable whether it can be at all justified to develop and test a contraceptive method that carries the risks of the person using it developing auto-immune diseases and the potential further risks associated with the treatment of such diseases. As Gustav J. V. Nossal, director of the Walter and Eliza Hall Institute of Medical Research in Melbourne, Australia, states:

> Treating autoimmune diseases necessitates abolishing or at least restraining the immune system. Immunosuppressive and anti-inflammatory drugs can achieve the desired effect, but such a blunderbuss approach suppresses not only the bad, antiself response but also all the desirable immune reactions. (Nossal 1993:27)

Gordon Ada, researcher at the Department of Immunology and Infectious Diseases, Johns Hopkins School of Hygiene and Public Health in Baltimore, has stated clearly that 'any vaccine to control human fertility that was shown to induce auto-immune disease during clinical testing or even at the post-registration stage would be considered unsafe for general use' (Ada 1990:573).

However, if severe adverse effects occur infrequently, they are unlikely to be detected until large numbers of users have been exposed to an immuno-contraceptive. Furthermore, as Ada and Griffin point out, it is extremely difficult to establish a causal relationship between the use of immune-mediated methods and adverse effects such as auto-immune diseases or allergies (Ada and Griffin 1991b:9).

Allergies Allergies or hypersensitivity reactions are exaggerated or inappropriate immune responses which can be triggered each time the body is exposed to an agent that stimulates the immune system. Allergic reactions after anti-disease immunizations range from sizeable but local inflammatory reactions at the injection site to generalized allergic reactions including rare, but potentially fatal, cases of shock.

In the Phase I (safety) trial of NII's anti-hCG contraceptive, 10 per cent of the women became allergic to the tetanus carrier after one of the three primary immunization doses (Talwar et al. 1990a:302). According to the trial report, 'out of 88 subjects who were immunized with different formulations of the hCG vaccine, 63 subjects did not have any complaints following first injection. The remaining 25 subjects (28 per cent) had minor complaints such as erythema [localized rash], pain at site of injection, fever, oedema, generalized rash, transient joint pain, nausea, muscle pain and giddiness' (Talwar et al. 1990a:305–6). Some of these adverse effects are clear signs of allergic reactions to the injection.

In HRP's Phase I trial, 2 out of 30 women (7 per cent) developed a hypersensitive reaction to the diphtheria toxoid carrier about forty-eight hours after the injection (Griffin 1988:187).

The allergic reactions observed in both trials occurred during the primary immunization schedule, before any booster injections. Moreover women considered to be at risk of developing allergies had been excluded from both trials. The incidence of allergic reactions in the HRP trial may be lower than that of the NII trial because HRP used only two primary injections. However, the HRP team stated that the incidence of allergic reactions observed in the trial 'would probably be regarded as an unacceptably high level when large-scale use of the vaccine is considered' (Griffin 1988:185).

The HRP team concluded the report of its Phase I clinical trial with a recommendation 'to screen all individuals with [a diphtheria skin] test before repeat vaccination' (Jones et al. 1988:1,297–8). But to exclude foreseeable severe allergic reactions, it would be necessary to test every woman before each injection, a measure that seems impractical in routine family planning settings. Immunologist Faye Schrater emphasizes that 'tests for hypersensitivity, or allergies, generally require a 48-hour period, thus adding another complication to [antifertility] vaccine use. That complication is necessary: although most allergic responses simply cause discomfort, in rare instances they can be fatal' (1992:44).

The use of tetanus or diphtheria carriers in anti-hCG immuno-contraceptives is a cause for concern. It is known, for instance, that 'immunization, particularly with diphtheria and tetanus toxoids, may result in increasingly severe local reactions' (Cohen 1987:674). In anti-disease vaccination programmes worldwide, the period between repeat immunization for diphtheria and tetanus has been gradually extended because of the risks associated with these vaccines. It is recommended to wait for 'at least three years' after the end of the primary immunization schedule before giving children a booster injection of tetanus or diphtheria. Moreover, because of the risks, adults usually receive a much lower dose than children (BNF 1995:495).

HRP is now proposing to use a different carrier instead of its diphtheria toxoid (HRP 1988:187). Vernon Stevens's optimistic review of the experimental possibilities concedes that 'today immunologists cannot assure that it is feasible to design [an antifertility] vaccine which will elicit high antibody levels without some degree of immediate or delayed hypersensitivity' (Stevens 1992:138).

Immune-complex diseases   The most frequent immune-complex-mediated tissue injuries are lesions around the site of injection and kidney damage. Such immune hazards are caused whenever antibody–antigen complexes are not sufficiently broken down and eliminated from our body through the usual waste elimination processes, but are deposited in smaller blood vessels where they cause inflammation.

With anti-disease vaccines, immune-complex formations can occur at each immunization or upon contact with the infectious agent. In the case of immuno-contraceptives, immune complexes may be formed whenever a person's body produces the natural counterpart of the administered antigen (for example, hCG with anti-hCG contraceptives). Theoretically, the higher the concentration of the antigen and/or the more persistently it is present in the body, the higher the risk of an immune-complex formation.

When the second wave of development of immuno-contraceptives started in the late 1970s, a basic recommendation was that 'to minimize the risk of immune complex disease, the target antigen should not be continuously present in the vaccine recipient but only intermittently and/or at low concentrations' (quoted in Ada and Griffin 1991a:xvi).

The use as antigens of non-pregnancy-associated reproductive hormones (FSH, GnRH) and of egg or sperm structures is expected to be the most risky in terms of immune-complex diseases. The beta-hCG contraceptives under development by the NII team and by the Population Council also run a comparatively high theoretical risk of causing immune-complex lesions in the pituitary gland due to their cross-reactions with the hormone LH produced continuously by the pituitary gland. NII and Population Council researchers claim that a long-term study of sixty rhesus monkeys provided a 'reassuring verdict on the lack of deleterious side-effects' (Talwar and Raghupathy 1989:98). HRP's David Griffin maintains, however, that 'the question of long-term immunopathological or other sequelae, if any, of cross-reactive immunity to hLH is still unresolved' (Griffin 1993:41).

Interaction with existing diseases   The adverse effects discussed above could occur in any healthy person who receives an immuno-contraceptive. Other effects could take place, however, if someone who already had a disorder of the immune system or another disease received an immuno-contraceptive. According to the 1989 HRP Symposium on Vaccines for Fertility Regulation, there are the following theoretical risks:

— Pre-existing allergies or auto-immune diseases (such as rheumatoid arthritis; could be exacerbated due to the 'general immunostimulation' inherent in the administration of antifertility vaccines.

— A person with a genetic predisposition towards immune disorders may develop auto-immune reactions or allergies for the first time.

— The risk of developing a chronic liver disease could be increased in persons with hepatitis B. This includes both people who have hepatitis B and those who are carriers (that is, persons who are infected with the hepatitis B virus, but do not show any symptoms). In many African and Asian countries, 5–10 per cent of the population are thought to be hepatitis B carriers (Nossal 1993:30).

— An HIV infection might progress quicker towards full-blown AIDS and might increase the risk of auto-immune diseases developing after administration of an immuno-contraceptive. In addition, the efficacy of the contraceptive could be decreased by HIV infection (Report 1991:289, 290).

These interactions are, as yet, hypothetical because only healthy, carefully screened people have been enrolled in clinical trials. At the 1989 HRP symposium, it was recommended that 'evidence of HIV infection be an exclusion criterion for Phase 1, 2 [and] 3 clinical trials of antifertility vaccines' (Report 1991:260). Immunologists reviewing *Vaccination against Pregnancy* have stated that, based on existing theoretical knowledge, it is not likely that administration of an immuno-contraceptive would speed up HIV's progression to AIDS – but it is certainly possible that HIV infection may interfere with the effectiveness of an immuno-contraceptive.[11]

A common feature of the conditions outlined above as theoretical risks is that it is extremely difficult, if not impossible, to screen them out so as to ensure that persons who have such conditions are not exposed to the effects of immunological birth control methods. It is not possible to test for a predisposition to an auto-immune disease or allergy in a routine family planning setting. Nor could most health care systems, particularly in Third World countries, afford the costs of HIV and hepatitis B tests before administering immuno-contraceptives.

Researchers of the NII, HRP and Population Council teams agreed in 1992 with women's health advocates that immuno-contraceptives 'would not be recommended for individuals at high risk of infection with HIV, or with other conditions which adversely affect the immune system' (HRP 1993:20). The current focus of public attention on AIDS as a viral disease has tended to obscure the social and economic causes underlying the spread of AIDS. It has also diverted attention from the fact that widespread diseases of poverty such as tuberculosis, malaria, schistosomiasis and other parasitic worm infections – as well as malnutrition – can all depress our immune system to varying degrees.

Effects on the foetus

Immuno-contraceptives combine three characteristics that significantly increase the probability of unintended pregnancy and the risks of foetal exposure to ongoing immune responses:

— a relatively long lag phase following primary immunization;
— the difficulty of predicting contraceptive failure, which may be due to individual variations in immune responses and a myriad of internal and external body factors;
— the impossibility of specifically 'switching off' ongoing contraceptive immune responses.

If a woman has sexual intercourse, unintended conception may occur either in women whose induced immune response is below the threshold level for contraception or in women who have reached the threshold antibody level but for whom the method has failed during the supposedly contraceptive phase without them being aware of it.

Depending on the reproductive component targeted by an immuno-contraceptive, the foetus could be damaged through a variety of mechanisms at various stages of pregnancy; potential problems outlined at the 1989 HRP meeting on antifertility vaccines included miscarriage, various visible malformations, and less apparent hormonal abnormalities. Some of these adverse effects, for example 'abnormalities of secondary sex characteristics, and development of possible neoplasia [cancer]' might manifest themselves only at puberty (Report 1991:289).[12]

During foetal development, the foetus's immune system begins to differentiate between its own body components and foreign agents. Any interference with the mother's immune system may affect the foetus as well. Without long-term animal studies, it is difficult to predict the possible effects of prenatal exposure on the child's immune system.

Additionally, breastfeeding transfers antibodies from the mother to the baby, so that the baby is protected for a certain period of time from diseases against which the mother has developed antibodies. Whether antibodies produced from immuno-contraceptives pass into the breast milk and what type of effects they may have on the baby are unknown.

Protection against sexually transmitted diseases and HIV infection

It often seems to be forgotten in contraceptive research and provision that sexual intercourse can lead to the transmission of sexually transmitted diseases (STDs) as well as to pregnancy. Instead, STDs and pregnancy tend to be considered by some researchers and policy makers as two separate issues requiring separate solutions.

Some contraceptives, in particular the condom (and to some extent withdrawal), can prevent pregnancy *and* provide protection against sexually transmitted diseases, a consideration that has become increasingly important with the spread of AIDS throughout the world. The action of other contraceptives may not affect STD transmission, whilst still others may actually increase the contraceptive user's susceptibility to STDs including HIV. For example, a large European study of the sexual partners of HIV-positive men found that women using an intra-uterine device (IUD) had the highest risk of contracting the virus (European Study Group 1989). Injectables and implants can also increase the risk of HIV (and hepatitis B) transmission if sterile equipment is not used.

The presence of HIV is just one change in the pattern of STD infection in the last two decades. Syphilis and gonorrhoea have been joined by new syndromes associated with the human herpes and papilloma viruses and the organism chlamydia. 'The second generation of sexually transmitted organisms are frequently more difficult to identify, treat and control. Moreover they cause serious complications which can result in chronic ill health, disability and death.' (HRP 1992:13) Chlamydia and gonorrhoea can lead to pelvic inflammatory disease (PID) if they spread to a woman's upper reproductive organs, a disease that frequently leads to infertility and chronic pain. Rates of PID are high in some developing countries. For example, each year 1–3 per cent of women of reproductive age in urban areas of sub-Saharan Africa develop PID. Village studies in India, Kenya and Uganda found rates of PID as high as 20 per cent (Jacobson 1991).

Thus in the 1990s, a pertinent question about any new contraceptive is whether it will protect users not only against pregnancy but also against STDs. Current research on immuno-contraceptives began in the early 1970s, when there was no awareness of HIV and AIDS. The WHO now forecasts that by the year 2000 a projected 80 per cent of HIV infections will be contracted by heterosexual intercourse – 90 per cent of them in developing countries (HRP 1992:14). One can only agree with HRP that:

> the AIDS pandemic has implications for contraceptive technology. Contraceptive choices, at the individual and programme level, will have to take the risk of prevalence of HIV infection into consideration. The need for dual protection against unwanted pregnancy and against STDs/HIV poses a challenge to contraceptive development. (HRP 1992:13, 14)

Immuno-contraceptives will not provide any protection against STDs whatsoever. In fact, besides the issue of the potential direct interaction with HIV infection (see p. 55) immuno-contraception may have important indirect consequences in increasing the incidence of STDs, particularly in developing countries.

The introduction of immuno-contraceptives could also cause setbacks to public health campaigns against the spread of HIV. Eka Esu-Williams, an immunologist heading the Society for Women and AIDS in Africa, expressed these fears at the 1992 HRP Meeting on Fertility Regulating Vaccines attended by women's health advocates and immuno-contraceptive researchers. She pointed out, first, that any injectable contraceptive is likely to contribute to the spread of HIV via improperly sterilized needles and, in the case of immuno-contraceptives, various blood tests, and, second, that the anticipated wide and intensive promotion of antifertility 'vaccines' would probably reverse any progress made in the endeavour to persuade men to use condoms to prevent the sexual transmission of HIV.

Contraceptive research decision makers, including those setting the priorities for development of new contraceptives, should support public health policies aimed at decreasing the spread of AIDS and other STDs. They should reconsider the continued development of any birth control method that risks contributing to the spread of STDs. This should be a priority worldwide, but it is especially important for contraceptive methods that are explicitly being developed for use in Third World countries, many of which already have high rates of AIDS and other STDs.

Service delivery requirements

The benefits and risks of any biotechnology depend not only on the technology-inherent characteristics but also on how this technology would be provided in practice. To ensure maximum effectiveness and minimum risks of immuno-contraceptives, the following measures would have to be taken by those providing the contraceptives:

— medical examinations should be made to ensure that people with certain diseases and pregnant women are not given the immuno-contraceptive;
— another contraceptive that does not interfere with the immuno-contraceptive must be provided for the lag and waning periods;
— antibody titres must be tested repeatedly to check whether the immune response is sufficient to be effective as a contraceptive or has started to wane, and whether current infections, malnutrition or stress have depressed the immune response;
— if the primary immunization schedule involves more than one injection, it must be followed rigorously, and booster immunizations must be given at the appropriate times;
— careful medical follow-up is required.

All these measures are critical because users will not be able to discontinue the effects of immuno-contraceptives at will.

Anticipated biomedical characteristics and risks

In conclusion, it is highly unlikely that immuno-contraceptives will ever be as effective as other 'systemic' contraceptives (that is, methods that interfere with complex body systems and whose increased risk – compared with those of barrier methods such as the condom – have to be justified by a high method effectiveness).

The question of the overall method effectiveness is not the only problem with immunological birth control methods. A woman using such contraceptives may find that it takes anything from a few weeks to a few months for the contraceptive to become effective and that she will not be able to tell when the effect is sufficient to prevent pregnancy or has worn off; she may find it will not work at all or, at the opposite extreme, that she has become permanently infertile. Moreover, the lag period (which will, however, probably be much shorter at subsequent 'booster' immunizations) means that another back-up contraceptive has to be used.

These are not simply technical flaws that can be solved by adjusting the formula of the immuno-contraceptive. They reflect the variability and lack of predictability of immune responses in general, factors that are bound to have an increasing impact on both the contraceptive and the adverse effects of immunological birth control methods as they move from being used in closely monitored trials with carefully selected people to the real life conditions of users.

Although less is known at present about the safety of immuno-contraceptives than about their effectiveness and reliability, concerns about risks of auto-immune diseases, allergic reactions, immune-complex formation and exacerbation of existing diseases cannot be lightly dismissed.

Foetal exposure seems likely because of the unpredictability of the duration of the contraceptive phase, the lag period and the gradual waning of the method's effectiveness, and the difficulty of 'switching off' ongoing immune responses. The effects on the foetus of exposure to an immuno-contraceptive are unknown but potentially very serious. Again, these flaws are inherent in any immune-mediated contraceptive.[13]

Even if immuno-contraceptives do not speed up the progression of HIV towards AIDS, they may well increase the spread of HIV and other STDs in indirect ways. On the other hand, AIDS and any other immuno-suppressive condition will further decrease the effectiveness of immunological contraceptives.

For safe use, repeated clinic visits would probably be necessary, even if a contraceptive required only one primary injection and was initially effective for one or two years. Before anyone was given an immuno-contraceptive, it would be necessary to test that they did not have any contraindications such as an existing auto-immune disease, allergy or any other disease that might be exacerbated by administration of an immuno-contraceptive or that might decrease its efficacy.

It is doubtful whether even optimal health care services could guarantee that immuno-contraceptives would be acceptably effective and safe for users. The ongoing privatization of health care services the world over may further increase their risks to those who cannot afford such health care. Can already overburdened health care systems in developing countries, which in some places are being dismantled as a result of structural adjustment programmes, be expected to implement all these measures?

Notes

1. Ada and Griffin 1991b:8
2. In Staines, Brostoff and James 1993:5.
3. Many men may also wish that more contraceptive options were available to them, but given the power differences between women and men and the effects of a pervasive patriarchy, their options and wishes are bound to be shaped by different considerations from those of women. In terms of reversible birth control methods, men's options are limited to condoms, which are often not taken seriously by health professionals or population controllers as means of birth control, and moreover have a tainted image as a means primarily to prevent the transmission of STDs.
4. I use the term user-centred because I believe there should be no difference between what is currently called a 'user perspective' of a contraceptive and a 'researchers' perspective'. Both the research community and interest negotiators should try to put themselves in the place of anticipated users and to assess the technology from their various perspectives with their differing views and in their differing contexts. In this publication, most of my assessment centres on women users, because it has been and still is harder for them to prevent outside control over their reproduction. Although two immuno-contraceptives are being designed for use in men, there is a gap concerning the criteria and theorizing from a male user perspective, partially because the condom is still the only reversible means of contraception for men but also because of power differences in most societies between women and men.
5. The reader should be aware that any risk assessment carries a subjective element, not the least because the vocabulary for expressing risks is very difficult and not always clear-cut. The use of the term 'may', for instance, does not necessarily mean that an adverse effect has occurred in human trials, but it does connote a definite theoretical possibility of that adverse effect occuring. Whether or not that theoretical risk is worth taking is a matter of judgement.
6. It is not yet clear what this antibody threshold level should be. It would depend on the type of immuno-contraceptive and the target antigen. It is still unknown to what extent a certain antibody titre can, in fact, be an indicator of contraceptive efficacy. For example, antibodies may have a high or low affinity for the antigen which determines how effectively an antibody binds with an

antigen. Although a low amount of high-affinity antibodies may be more effective than a high amount of low-affinity antibodies, most studies have not distinguished between them (Talwar et al. 1990b:585). Only Phase II efficacy trials could give an initial indication of the correspondence between antibody levels and contraceptive effect. The NII Phase II trials seem to confirm the theoretical assessment of 50 nanograms per millilitre of blood for anti-hCG immunization (although this figure too does not distinguish between high- and low-affinity antibodies) (Talwar et al. 1994a).

7. The NII team also plans to develop a neem-based product to be injected into men's sperm duct (*vas deferens*) as a 'reversible method for male fertility control' (Talwar 1994a:3).

8. The mean value is calculated by adding all the durations of effectiveness together in a particular trial or those of a subgroup in a trial and dividing by the number of women in that trial or subgroup.

9. The 1993 version of *Vaccination against Pregnancy* (see Chapter 7) stated that in NII's Phase II trial there was an indication of a 'more than four-fold variation between individuals' from six months to over two years (Richter 1993: 32). This statement was based on a confusing preliminary report of the Phase II trials which was similarly misinterpreted by others (e.g. HRP 1993:21). The final report, however, made it clear that the women in the trial had been given anything from none to six or more boosters – to yield a contraceptive duration ranging from less than three months to more than thirty months in those 80 per cent of women who reached the contraceptive threshold (20 per cent did not) (Talwar et al. 1994a:8,533). Thus there was no standard regimen for all trial participants of, for instance, the primary immunization course and one booster injection.

10. HRP subsequently found 'adverse results' in mice when they attempted to counteract anti-hCG immunization through progesterone replacement (HRP 1994a:128).

11. Jeff Spieler from USAID told women's health advocates at a 1992 HRP meeting in Geneva that his literature search did not show any risk of the exacerbation of HIV infections after anti-disease vaccinations. However, he also stated that research on the interaction between HIV and anti-disease vaccination is as yet scant.

12. This occurred with foetal exposure to DES (diethylstilbestrol), a synthetic oestrogen that was widely prescribed in the 1960s to prevent miscarriage. It was later found to have induced rare forms of vaginal and cervical cancer (clear cell carcinoma) in daughters of women who had been prescribed the drug. These drug-induced cancers develop around twenty years after foetal exposure. For other problems caused by DES, see Chetley 1995:226–8)

13. Some commentators maintain that if a woman becomes pregnant while using a contraceptive, she can just have an abortion. After all, she was presumably taking the contraceptive because she did not want a child. Such a comment does not consider whether a woman has access to safe, legal abortion services, or that the wish not to have a child when one is not pregnant does not automatically translate into a wish not to have a child when one *is* pregnant. I am in favour of abortion as a woman's right, but oppose it as a duty.

4

Social risks of immuno-contraceptives

It is important that any discussion of reproductive technology include an analysis of reproductive freedom which focuses on individual control and protests against coercion (either from other individuals or from social forces generally) over central aspects of reproductive life. The norm of reproductive freedom should protect individuals' control over their own sexuality, sexual activity, and childbearing and ensure adequate resources for childrearing. (Susan Sherwin, philosopher, Dalhousie University, Halifax, 1989)[1]

It is essential to assess technology within an actual social context … In studies of women and technology, this means recognizing the conflicts and contradictions in women's lives. (Janine Morgall on technology assessment, 1993)[2]

Potential for abuse

The principal aim of a contraceptive, obviously, is to prevent pregnancy temporarily. According to the Oxford English Dictionary, the word 'contraception' came into use at the beginning of this century meaning 'prevention of uterine conception'; it was a combination of the words 'contra' and 'conception'. Earlier names for a contraceptive had been 'anticonception' and 'contraceptic'. It is thus commonly assumed that contraceptives are used to prevent 'unwanted' pregnancies – but unwanted by whom?

While modern reliable contraceptives have made it easier for women (and men) not to have children at a particular period in time but still to have sexual intercourse, they have also made it easier for powerful social actors to attempt to control certain women's fertility so that they do not have children or 'too many' children, irrespective of the wishes of the individual woman or man or family.

Rather than focusing on the prevention of unwanted pregnancies, a feminist assessment of the benefit of a particular contraceptive should focus on its contribution to people's, particularly women's, *reproductive self-determination*.[3] In doing so, it can be seen that high method effectiveness is not automatically of benefit; indeed, it may be an obstacle to the empowerment of marginalized people, particularly if a method has 'abuse built into its design' (Hartmann 1995:213). Such abuse is not necessarily deliberate or planned, but can be a consequence of the various assumptions that have guided the development of contraceptive technology over the past forty years or so.

Contraceptive abuse is more than forcing people to use a certain contraceptive against their will or administering it without their knowledge. Argentinian women's health advocate Mabel Bianco has defined abuse of birth control methods as 'any instance in which a method is imposed upon a person or in which a person is induced to use a particular method in a way that the decision is not the result of free and informed choice' (HRP and SERG 1994:4).

Abuse can thus take place in a number of ways, including:

— administering a contraceptive without a woman's consent or knowledge;
— implementing sanctions against nonusers;
— providing financial incentives to encourage 'preference' for a particular method;
— refusing to remove a contraceptive when the user so requests;
— giving biased information about a method, such as emphasizing its effectiveness while playing down or not mentioning its adverse effects so as to convince specific groups of people to use a particular contraceptive and to dissuade them from using others.

Much of the biomedical and family planning literature attributes such abuse (when it is acknowledged) to 'overzealous' providers (without considering what has made them 'overzealous'), portraying the contraceptives themselves as neutral technologies. But as social scientist Judy Wajcman points out, it is social forces that determine what contraceptive technologies are available – and available birth control methods can in turn shape our life 'choices' in both technical and social terms (Wajcman 1994:154).

The aggregate of three aspects of a contraceptive technology is at the core of assessing a contraceptive's *potential for abuse*:

— the duration of the anti-fertility effect;
— the possibility or impossibility of the user stopping this effect when she or he wishes (degree of user control);
— the dosage form or device (that is, whether the contraceptive is a barrier device, a pill, an injection, implant or an intrauterine device).

Table 4.1 provides an overview of the duration, degree of user control and delivery systems of several types of contraceptive and thus assesses their potential for abuse.

Any method whose effects last for a long time and that the user cannot stop when she or he wishes lends itself to abuse. It is difficult, for example, to imagine abuse of a condom. As Maggie Helwig of the Coalition for East Timor said in concluding her description of soldiers

Table 4.1 A comparison of abuse potential of contraceptives

Method	Duration of effectiveness	Possibility of stopping effect at will	Dosage form/device	Abuse potential
Barrier methods	During intercourse	Can be removed by user at any time	Condoms, vaginal barrier (±spermicide)	None
Oral hormonal contraceptives	1 day	Can be stopped by women at any time	Oral	Low
Vaginal rings with hormones[a]	3 or 6 months	Can be taken out by women at any time	Vaginal slow-release system	Low
Injectable hormonal contraceptives	1, 2 or 3 months[b]	Women must wait until the effect wears off	Injection	High
Hormonal implants	5 years	Can be removed at any time, but only by specially trained health workers	Six capsules under the skin. Minor surgery needed for insertion and removal	High
IUDs	1 to 8 years	Can be removed at any time, but only by specially trained health workers[c]	Intrauterine device, inserted and removed through the cervix	High
Immunological contraceptives[a]	Potentially 1 year to lifelong	Women must wait until the effect wears off	Some injection, some oral	High to very high

Notes: [a] Under development [b] Return of fertility may be delayed after effect wears off [c] In emergencies or in cases where removal has been refused, IUDs with a string have sometimes been removed by women themselves rounding up women from East Timorese villages so that Norplant® could be inserted, 'You can't force a guy to use a condom at gunpoint.' Of the various hormonal methods, the Pill has little potential for abuse because the action of a single tablet lasts for one day only and a woman can stop taking it at any time. The effects of injectable hormonal contraceptives, however, last for one to three months and cannot be reversed during this time; injectables therefore have a higher abuse potential. Hormone-releasing vaginal rings currently under development will be effective for three or six months, but a woman can remove them by herself any time she wants. So although injectables and vaginal rings have a comparable duration of action, they each have a profoundly different abuse potential because of a different degree of user control.

The implant Norplant® acts for five years and is regarded by its developer, the Population Council, as being an intermediate between contraception and sterilization. When it was introduced by the Population Council, women were told that its advantage over hormonal injectables was that it could be removed at any time and its effects halted. In theory this is true; in practice, Norplant® needs the cooperation of a health care provider to remove it surgically. In demographically oriented family planning programmes aimed at bringing birth rates down, women have often had difficulties in getting the implant removed before the end of its five-year contraceptive duration. (Mintzes et al. 1993).

A Population Council study entitled 'Service delivery systems and quality of care in the implementation of Norplant in Indonesia' (Ward et al. 1990:30) is revealing in this respect.[4] An Indonesian family planning manager explained that 'people are told that it has to last five years, they give their word … and rural people don't go back on their word. If they request removal, they are reminded that they gave their word.' Nearly one-third of the women interviewed for this study said that they did not dare even to request removal because they feared their request would be refused.

Abuse of birth control methods does not only occur in demographically driven family planning programmes. Immediately after Norplant® was introduced on the US market in 1990, an editorial in the *Philadelphia Inquirer* advocated Norplant® as a 'tool in the fight against black poverty' (quoted in Hartmann 1995:211). Within just two years, some twenty legislative bills, amendments and welfare reforms regarding its use for specific social groups had been proposed in thirteen states across the country. One bill in Mississippi advocated compulsory insertion of Norplant® for women with four or more children receiving state assistance (even if this assistance was only medical aid, housing assistance or disability benefits). Most of these proposals were rejected (Alan Guttmacher Institute 1992). There have been specific court orders, however, for Norplant® to be inserted in women convicted of drug or other criminal offences. The characteristic of Norplant® as 'temporary sterilization' has encouraged the resurgence of debates in which African, Hispanic, Native American and poor white women are singled out as undesirable 'breeders' (see Hartmann 1995:254–6; Davis 1990).

The type of contraceptive device is the third important technology-inherent feature of a method's potential for abuse. It harbours three potential consequences:

— the ease with which a method can be distributed on a mass scale;
— the degree to which it can be given without people's knowledge;
— the ease with which people can be persuaded of its 'advantages'.

The more easily a method can be administered on a mass scale and given without people's knowledge, and the more easily its 'acceptability' can be engineered to increase its use, the higher the risk that it will be abused. Of course, abuse cannot be predicted mathematically. The actual abuse of a particular contraceptive depends on its technology-inherent features *and* the specific sociopolitical and cultural context in which it will be introduced. However, in the light of women's historical and present-day experiences with birth control methods, a guiding principle in forecasting a method's abuse could be: hope for the best – but be prepared for the worst.

The abuse potential of antifertility 'vaccines'

Some contraceptive developers see antifertility 'vaccines' as 'an unprecedented effective instrument for demographic control' (Mitchison 1991:249). Because of their technology-inherent features and given the history of population control, I see them as contraceptives with an unprecedented potential for abuse.

Their duration of action may be anything from one year in the case of anti-hCG versions to potentially lifelong in the case of those directed against egg or sperm structures. The effect of immuno-contraceptives cannot at present be discontinued safely, if at all, by a user. The delivery system of an immuno-contraceptive will depend on its type and formulation; some may be injections, others pills or liquids.

The potential of the uses and effects of these new technologies – and thus their potential for abuse – can be glimpsed by considering reactions to research into the immunological birth control of animals. 'New animal vaccines spread like a disease', announced a *New York Times* headline to an article that described a proposal to incorporate sperm antigens into live viruses hidden in bait which would be distributed among wild animals to eat. Once consumed by an animal these viruses, and thus the sperm antigens, would be transmitted to other animals of the species, just like infectious diseases. The journalist was enthusiastic about the prospect:

> Biologists say that [the] new vaccines ... will provide a humane method for drastically reducing [the] population of rabbits in Australia, rats in Indonesia, white tailed deer in the United States and other rapidly multiplying species that threaten the environment ... Since the vaccines work by immunizing a female against the male's sperm, the same principle should be effective as a contraceptive in humans ... American research leaders believe that within the next decade an oral contraceptive vaccine could be available for test on human subjects. A single dose, it is hoped, could confer temporary infertility for years ... the method could make contraception far more accessible to residents of poor countries. (Browne 1991:C1, C6)

Two US researchers, CONRAD-supported John C. Herr at the University of Virginia and Roy Curtis working at Washington University in St Louis, are already collaborating on an oral anti-sperm contraceptive for women that incorporates the antigen into altered salmonella bacteria, an organism that in its natural form can cause fatal food poisoning (Browne 1991:C7).

Talwar, too, is advocating the development of 'live recombinant vaccines' to lower the costs and increase the efficacy of his anti-hCG immuno-contraceptive (Talwar 1994b:702). As mentioned in Chapter 1, the National Institute of Immunology in India has tested its anti-hCG formula inserted into a live vaccinia virus as a treatment for certain types of lung cancer and is planning to expand that trial (Talwar 1994b:702). As a next step, the NII team plans to incorporate beta-hCG into fowlpox virus – rather than vaccinia virus – claiming that this vector would be 'without risk in immuno-compromised subjects' (Talwar 1994a:2).[5] Unlike the genetically engineered rabbit 'vaccines' described in the *New York Times*, products for humans are not intended to be transmissible between people. But no one knows to what extent they may in fact spread, albeit unintentionally. The 1989 HRP report on antifertility vaccines calls for an evaluation of 'the possibility of vector [i.e. those bacteria or viruses incorporating microbial or reproductive antigens] transmission between individuals' (Report 1991:260).[6] The creation of a technology whose antifertility effect could accidentally be transmitted between people would radically change the perception of 'user control'.

Whether the delivery system of immuno-contraceptives will be injections, pills or liquids, they could certainly be administered much more easily on a mass scale – with or without a person's knowledge or consent – than existing birth control technologies. Even such relatively complicated procedures as IUD insertion and sterilization have been carried out without women's knowledge, for example immediately after a woman has given birth, when she may not be fully aware of what is happening around her or when a woman is under general anaesthesia or (in the case of IUDs) during a routine gynaecological check-up. For immuno-contraceptives, there would be no need to wait for such occasions; they could easily be given at any time when a woman or man requested or agreed to an injection or pill for the treatment or prevention of a disease.

Many women's experiences with the introduction of new contraceptives illustrate that the grey area of 'disinformed consent' – or 'promotion' – is one of the most significant areas of contraceptive abuse.[7]

That immunological birth control methods can be promoted as 'vaccines' has been cited as one of their major advantages since the beginning of the second wave of research (see Chapter 5). According to a consultation document issued by HRP in 1978: 'immunization as a prophylactic measure is now so widely accepted that it has been suggested that one method of fertility regulation which might have wide appeal as well as great ease of service delivery would be an anti-fertility vaccine' (HRP 1978:360). This opinion was reaffirmed in 1987 by USAID biomedical adviser Jeff Spieler:

> Fertility regulating vaccines should be well accepted by users; given the general popularity of immunization there would be compelling advantages for service delivery because vaccines … could be administered by paramedical and non-professional personnel, and could be integrated not only with family planning services but with other health care programmes as well. (Spieler 1987:779)

The description of immuno-contraceptives as 'vaccines' is anything but informative. It obscures the complete novelty of this birth control technology, projecting instead a reassurance that because some vaccines have been useful in disease prevention, antifertility vaccines must be beneficial too.

If a contraceptive's *potential for abuse* is defined as 'technology-inherent features which increase the likelihood of uninformed, disinformed and coercive administration of a birth control method' (Richter 1994a:216; see also note 19, Chapter 7), the abuse potential of antifertility 'vaccines' puts them way beyond any form of social means of preventing or containing such abuse. They will be relatively long-acting, cannot be discontinued by the users at will, can easily be mass-administered – with and without the knowledge of users. The explicit intention of using the image and comparison of anti-disease vaccines to promote immuno-contraceptives should be a further cause for concern about uninformed provision on a mass scale.

Confidentiality

> In developing countries, men often oppose the use of contraceptives. This vaccine would allow women to use a birth control method without being detected. (Rosemarie Thau, Research Coordinator of the Population Council, 1993)[8]

The argument that some women may want to use long-acting contraceptive methods has been put forward in recent debates about immuno-contraceptives not only by the research community, but also by women working as consultants in family planning and health care programmes.

Indeed, it is a sad fact that various forces and influences often put women worldwide – not only in countries of the South – in difficult positions concerning their fertility and reproductive decision making. Women's partners, families, religious leaders and now the New Right may object to women using contraception or having an abortion. In such difficult circumstances, some women may indeed decide to resort to longer-acting methods that cannot be (or are difficult to be) detected by others (even though these tend to have more adverse side effects) rather than use no contraceptives at all.

It is one thing if a woman is given the opportunity to decide freely on the use of a long-acting contraceptive after being given complete and understandable information not only on this but also on all alternative methods. Yet how many women 'prefer', for example, hormonal injectables simply because this long-acting method has been one of the few contraceptives made available to them, often with inadequate information about their long-term risks and promoted with reference to the high status enjoyed by injections in many Third World countries?

Although many women are in the unfortunate position of being forced to make bad choices because of their individual circumstances, this is not necessarily an argument for the continued development of antifertility 'vaccines'. A women-centred technology assessment has to make sometimes painful choices between 'what is best for the individual woman and, in a larger context, what is best for women in general' because 'these two issues may be in conflict' (Morgall 1993:204). Any perceived increase in space for some women's reproductive self-determination due to the possibility of concealing a contraceptive device (or practice) has to be assessed against a probable accompanying decrease of decision-making space for these and other women because of the birth control technology's potential for abuse.

Given the availability of other birth control methods that women can hide, instead of increasing the range of long-acting contraceptives through the addition of immunological birth control methods it would seem to be more of a priority to make sure that any woman using a long-acting method does so in full knowledge and understanding of its risks and with good medical back-up. For those concerned with in-creasing women's ability and potential for decision making, there is no shortcut to persuading men and various social actors that act against women's reproductive and sexual self-determination and overall self-determination that women should be able to use contraceptives of their own choice – while, at the same time, dissuading demographically minded, or otherwise prejudiced, providers and policy makers from pushing particular birth control methods on women.

At issue is more than whether some women in some contexts may want another long-acting contraceptive method. Even if they did, would they want antifertility 'vaccines' if they were informed that the method uses their immune system, that its action will last for at least a year, whose action they cannot stop, which can easily be given without their knowledge, and which has potentially serious side effects?

Women's complaints about the fact that it is not possible to stop the action of the injectable hormonal contraceptive in cases of adverse effects and the number of women worldwide who ask for IUD or Norplant® removal before the end of the contraceptive period seem to indicate that women may not find such a contraceptive attractive. In fact, recent events described below in several countries indicate that women may strongly object to the introduction of a contraceptive that would make it difficult if not impossible for them to trust family planning and health care providers.

Immunization backlash

The fact that it is administered by injection makes it easy to confuse, in-tentionally or unintentionally, with other preventive or curative injections … The advantage of being long lasting will be a problem instead of an advantage if the [antifertility] vaccine is given without the woman's informed consent. (Concepcion et al. 1991:239)

Experts invited to elaborate on the Social Aspects Related to the Intro-duction of a Birth Control Vaccine at the 1989 HRP symposium warned that the 'abuse of the birth control vaccine would not only harm family planning in general but it could also have negative consequences for public attitudes to other vaccines and to the health care system in general' (Concepcion et al. 1991:240).

This has happened faster than any of the 1989 symposium partici-pants, myself included, would have believed. At the 1992 HRP meeting between women's health advocates and scientists, reports were presented of Indian women refusing disease immunization because of rumours about sterilizing vaccines (Bang 1992).

By 1995, further reports of women's concerns at the possible administration of contraceptives in the guise of tetanus vaccines had come from Peru, Tanzania, Mexico, the Philippines and Nicaragua. Although women's health and rights advocates had forecast that the mere existence of antifertility 'vaccines' would make it impossible for women and men to trust family planning and health care systems once the method was available (for example, Richter 1994a:222, 227), the reasons behind this resistance to tetanus immunization is more complex.

One reason is that a number of countries have stepped up their tetanus immunization activities. In many developing countries, women are immunized against tetanus primarily during pregnancy. This is in order to prevent dangers for the woman and the newborn child if the umbilical cord is cut in unsterile conditions. In countries with a wide immunization coverage, tetanus immunization has also been carried out during infancy and childhood. A new WHO schedule recommended an increased focus on young women aged between fifteen and twenty years (Galazka 1993:10, 15–17), which has meant increasing immunization activity. In some countries, people's queries about what young schoolgirls (but no boys) were now being immunized for *en masse* coincided with information circulating about the existence of antifertility 'vaccines'.

In some instances, those spreading this information were not necessarily concerned about women's reproductive self-determination. In southern Tanzania, for example, a Catholic priest warned in February 1994 about abortion- and sterility-causing tetanus vaccines. The Kishuaheli-language newspaper article that spread his warning ended with the invocation, 'Oh God, save us from the tetanus vaccine' (*Mwenge* 1994). A month later, an English-language newspaper article in Tanzania warned about 'the dawn of a new form of abortion', namely anti-hCG vaccines that could 'cause abortion in any cycle in which the ovum is fertilized; ergo up to 12 abortions per year'. Author Lawrence Roberge continued, 'this clearly conflicts with biblical and church teaching on the sanctity of life and God's abhorrence of abortion', concluding, 'the church must not be lulled into complacency by world market acceptance; rather Christians must develop an appropriate response to this abortion vaccine' (Roberge 1994). Catholic authorities went on to prohibit tetanus vaccination in more than ten of their hospitals in southern Tanzania. The Tanzanian Ministry of Health had to approach WHO and UNICEF offices in Dar es Salaam for a statement that the tetanus vaccines in question had not been tampered with (Sangi 1994).

In the Philippines, rumours about contraceptive vaccines first hampered tetanus vaccinations during the First National Immunization Day in 1993 in the southern area of Mindanao. A year later, the impact of the rumours was observed nationwide (Health Alert 1994). By the end of 1994, the origins of these rumours became more concrete. In December, an anti-abortion group, Pro-Life Philippines, claimed that its Mexican counterpart organization had found evidence of hCG in tetanus vials (Symbahayan Commission 1995).[9] In February 1995, newspaper and radio reports in the Philippines warned that the tetanus vaccines distributed during the imminent Third National Immunization Day would also contain an abortifacient. The organization Pro-Life Philippines circulated an article by Roberge obtained from *Human Life International Reports*, a publication from the US headquarters of Pro-Life. In March 1995, Pro-Life Philippines obtained a Manila court order restraining any further administration of tetanus vaccines until the Ministry of Health had tested the vials and issued a reassuring statement. Yet many women remained suspicious. According to long-standing Filippino health activist Michael Tan, 'the numbers of women getting tetanus vaccines this year have dropped. Since mothers are avoiding the health centers, immunization against other childhood diseases such as measles has also dropped' (Tan 1995).

These examples of how resistance to immunization for fear of being immunosterilized came about illustrates the paradox that usually opposing interests – those who are against abortion such as the 'pro-life' movement and who tend in practice to oppose women's reproductive rights and self-determination as well, and those who are concerned about women's health and self-determination (the motivation for this book) – are both working to stop antifertility 'vaccines' (particularly, on the part of those opposed to abortion, the anti-hCG 'vaccine'). The motivation of each interest is obviously different, although such differences can be confused by outside observers.

Those concerned with women's reproductive rights are against the introduction of antifertility 'vaccines' not because some of them may act after conception but because of concerns over their impact on users' integrity and on the health and well-being of both users and their potential future offspring. When groups such as the Women's Global Network for Reproductive Rights have been sent vials from a particular country and asked to check whether the immuno-contraceptive is present, they have tried not only to do this but also to ascertain whether or not clinical trials are taking place or planned in the country as a possible further reassurance – no trials would suggest that the vial would not contain the contraceptive; there are no trials taking place in the Philippines, for instance. They have also tried to point out that, worrying as the antifertility 'vaccines' are, they are still at the research-and-development stage where close monitoring of trial subjects is needed, for instance by drawing and analysing blood at regular intervals in order to publish the results of the trial, results that are unlikely to be obtained from administering antifertility 'vaccines' indiscriminately to schoolgirls.

Once immunological birth control methods are on the market, however, groups such as the Women's Global Network will not be able to give such additional reassurances because the antifertility 'vaccine' could be clandestinely administered in any country. Ministries of Health will be ever busy with vial testing ... but who will believe them?

Anticipated social risks

As we have seen, three technology-inherent features of antifertility 'vaccines' are at the root of several social risks of the method. Their long duration of action, the impossibility for the user to simply 'switch off' their action, and the fact that they can easily be administered on a mass scale, even without the knowledge of the recipient, all make for an unprecedented potential for abuse. The method's potential benefit of allowing women to conceal their use of it from others who may oppose their use of contraception has to be assessed against the loss of spaces for reproductive self-determination for women the world over and cannot be considered separately from their biomedical risks and wider issues concerning the struggle to ensure women's voluntary and fully informed access to means of fertility regulation.

Immuno-contraceptives' high potential for abuse and the impossibility of the ordinary person being able to distinguish them from other injections or pills mean that their introduction would probably create a world in which women and men, particularly the underprivileged, would be able to trust neither immunization in general, nor the health services, nor the best possible family planning services.

Notes

1. Sherwin 1989:261.
2. Morgall 1993:201.
3. There is insufficient space here to discuss the complexities of the various terms *reproductive rights*, *reproductive self-determination*, *reproductive freedom* or *autonomy* and *reproductive choices*, or the various debates within women's movements about these terms. In the assessment of reproductive technologies, I prefer to use the term 'reproductive self-determination' because it allows social, political and technical aspects of human reproduction to be considered. For a discussion of the various terms and meanings, see WGNRR 1993. A resolution of the Sixth International Women's and Health Meeting, held in Manila in 1990, was that 'Reproductive rights means women's right to decide whether, when and how to have children – regardless of nationality, race, class, age, religion, ethnicity, disability, sexuality and marital status – in the social, economic and political conditions that make such decisions possible' (Philippines Organizing Committee 1991).
4. Betsy Hartmann first publicized the findings of the Ward Report, which was initially confidential. See Hartmann 1991 for a summary of these and other findings of the Ward Report.

5. According to Talwar, the 'fowl pox virus is species restricted and does not multiply in humans' (Talwar 1994a:2).

6. So far as I am aware, one of the transmissible animal sterilizing 'vaccines' is already being tested on an offshore Australian island. The virus used is a myxoma virus, an imported virus originally intended to kill off the rabbits that have eaten up much of Australia's vegetation since they were introduced to the continent some two centuries earlier and did not encounter any 'natural' predators. There is insufficient space in this publication to go into the details of how researchers can (or cannot) ensure that such 'vaccine' viruses will not mutate and will not cross to other species, or how they plan to prevent the virus from spreading to other parts of the world where there is not considered to be a rabbit problem. For related issues, see McNally 1994.

7. Norplant®, a method that needs minor surgery for both insertion and removal, was (and is still being) promoted as a 'convenient' method. By July 1993, 400 women were seeking damages from the US distributor of Norplant®, Wyeth-Ayerst, because of difficulties in removing the implant (Lewin 1994). The promotion of Norplant® occurred despite repeated warnings that the difficulties of removal were a technology-inherent problem of contraceptive implants (e.g. Mintzes et al. 1993).

8. Quoted in Barriklow 1993:29.

9. What had happened was that in October 1994, during the national vaccination campaign, parents of schoolgirls had asked the Mexican Ministry to test tetanus vials for the presence of hCG.

5

Advantages of immuno-contraceptives from researchers' perspectives

The stage is now set for the final steps in the development of a truly novel preparation which comes close to the ideal family planning method in that: it is free of effects on the menstrual cycle; it is devoid of conventional pharmacological action; it has a long duration of efficacy, perhaps following a single injection; it is independent of a continuous conscious act by the user for its efficacy; it is easy to administer by a procedure already associated with positive health benefits; and it is likely to be inexpensive when produced in large number of doses. (David Griffin and Warren Jones, HRP research team 1991)[1]

Chapters 3 and 4 have outlined some of the biomedical and social risks that immuno-contraceptives carry from a user-centred perspective. I now consider the benefits or advantages as claimed by those developing immuno-contraceptives, try to shed some light on why certain characteristics may be perceived as advantages, and assess whether these would also be perceived as advantages from the perspective of a person who would be using the method.

Throughout the scientific literature on the topic of immuno-contraceptives, five 'theoretical advantages' are cited. These were outlined in a briefing for participants at the 1989 HRP Symposium on Assessing the Safety and Efficacy of Vaccines to Regulate Fertility as follows:

— 'absence of pharmacological activity'
— 'long duration of effect from single injection or course of immunization'
— 'absence of user failure risk'
— 'administration by a method with high level of acceptability'
— 'low cost' (Ada and Griffin 1991c:18).

Absence of pharmacological activity

Given that this new class of birth control methods relies on the immune system, it is at first glance puzzling as to why a lack of 'pharmacological activity' – which one would not expect anyway – should be cited as an advantage. By 'pharmacological activity', however, researchers are primarily referring to 'menstrual disturbances and other hormone dependent side-effects' (HRP 1990:27). Thus any absence of pharmacological activity with an immuno-contraceptive is relevant if it is compared with hormonal contraceptives only and not with other methods.

In effect, the emphasis on 'menstrual disturbances' restricts the comparison still further to those hormonal contraceptives that tend to cause intermittent bleeding between periods, no periods at all, or excessive bleeding during periods; this is particularly the case with injectables and implants. As there are a variety of ways to prevent pregnancy, including barrier methods, that do not have pharmacological effects, the absence of menstrual and hormonal disturbances cannot automatically be an argument in favour of developing a new and potentially risky contraceptive.

In fact, those immuno-contraceptives that are directed against the hormones GnRH and FSH and those that lead to cross-reactions against LH, such as the formulas of the NII and the Population Council, are likely to have hormone-dependent short- or long-term side effects. In contrast to hormonal contraceptives, immuno-contraceptives targeting reproductive hormones carry additional risks from their reliance on the immune system (see Chapters 3 and 6).

Menstrual disturbances are not usually seen as the most important medical risk of hormonal contraceptives.[2] Why a lack of pharmacological activity, in particular a lack of menstrual disturbances, should be highlighted as a major advantage of immunological birth control methods seems to be connected with women's responses in the past to such activity. As is well known in family planning circles, the adverse effects of hormonal contraceptives, in particular menstrual disturbances, are a leading cause of women stopping using them, that is asking for removal of implants long before the end of their five-year duration or not going back for another injection (Snow 1994:241).

A contraceptive promising a lack of menstrual disturbances and other adverse effects that are perceptible and possibly disturbing to women could therefore be seen from a demographic standpoint as indicating the prospect of increasing 'continuation rates' and increasing the numbers of 'contraceptive acceptors'.[3] The introduction to HRP's report on the Phase I trial of its anti-hCG contraceptive, for example, stressed that a lack of 'discernible alterations in the menstrual cycle' would make it a 'highly acceptable birth control strategy' (Jones et al. 1988:1,295).

From a user-centred perspective, adverse effects are not a matter of 'acceptance' or 'continuation rates' but of health, well-being and general impact on lives. From that perspective, the research community should explain why exchanging the adverse effects caused by interfering with the hormonal system for the potential adverse effects caused by interfering with the immune (and potentially hormonal) system would be an advantage.

Long duration of effect

The long-lasting action of a contraceptive may be seen as an advantage by some women, for example, those who do not like to or are concerned that they will not remember to take the Pill every day, or to insert a diaphragm before sexual intercourse or to ask a partner to put on a condom during intercourse, for whatever reason.[4]

But as indicated in Chapters 3 and 4, any advantage of an increased duration of action has to be measured against a potential increase of adverse effects for users and their future children, and loss of space for reproductive decision making resulting from the difficulties, if not impossibilities, of 'switching off' immune reactions.

From a user-centred perspective, it is also a concern that some developers of immuno-contraceptives believe that long-lasting action would make continuous medical back-up unnecessary. (This has also been cited in the case of Norplant®.) Rosemarie Thau of the Population Council, for example, praises contraceptive vaccines for 'having little need for clinical support' (Mauck and Thau 1990:731), whilst David Griffin and Warren Jones assert that:

> The successful development of safe, effective and acceptable vaccines ... will be particularly beneficial to both users and providers of family planning methods in those countries where the storage and distribution of drugs and devices is difficult and access to health services is limited. (Griffin and Jones 1991:189)

In fact, long-acting methods that are not reversible at the discretion of the user require more careful delivery, pre-provision counselling and medical back-up than user-controlled, fully reversible, shorter-acting methods. As longer action has long been regarded as an advantage, however, within a demographic framework, increasing the duration of action of a contraceptive has been a priority in contraceptive research.

Absence of user failure risk

> One of the biggest problems in developing countries is that most contraceptives, like the pill or any of the barrier devices, require too much attention. So people don't use them ... One of the big advantages of a vaccine is that people won't have to do anything – aside from taking the initial injections. (Vernon Stevens, interviewed by a UNDP journalist)[5]

Lack of 'user failure' risk is a term that permeates the scientific literature as a much-sought-after goal of contraceptive development. As outlined in Chapter 3, the reasons why a method's effectiveness in real life situations may be less than its theoretical effectiveness under (so far as is possible) ideal and controlled circumstances are manifold.

Calling this difference 'user failure' puts the blame squarely on the user, absolving both providers and developers. It projects an image of contraceptive users as being 'ignorant' or incapable, closing off exploration of other potential reasons – and possible solutions. Many 'failures' of the Pill, for instance, could be a conscious decision to stop taking the oral contraceptive method because of its adverse effects. The provision and back-up service associated with a contraceptive can influence the failure rate as well; for instance, if service providers do not inform users properly and sufficiently about how to use the method correctly, or if users cannot obtain sufficient supplies of a method such as the condom, Pill or hormonal injections in time for continuous use, or if the method provided is of bad quality. In a user-centred framework of contraceptive research, the research and funding community would attempt to elucidate the reasons for the differences between theoretical and actual effectiveness rates, including an assessment of their own assumptions.

Any efforts to make contraceptives easier and more convenient to use would be welcome. But there is a profound difference between regarding those people a particular contraceptive technology is intended for as human beings active in a multitude of life circumstances, and perceiving them as passive, as 'failing' users. A perception of them as active would not have postponed improvements in condom technology and provision until the advent of AIDS.

In the case of immuno-contraceptives, a preoccupation with preventing 'user failure' seems to have led to the implicit acceptance of the creation of a class of contraceptives with a relatively high rate of 'method failure'.

Administration by a method with a high level of acceptability

By virtue of its mechanism of action, its lack of effect on the menstrual cycle, its long duration of efficacy, and the positive health benefits perceived to be associated with other forms of vaccinations, for example against infectious diseases, it is anticipated that an anti-hCG vaccine will find wide acceptance as a new method of family planning. (Griffin and Jones 1991)

From the beginning of the second wave of research into immuno-contraceptives, many researchers have regarded a major advantage of antifertility 'vaccines' to be 'the acceptability of vaccines in general [which] might facilitate their introduction' (Mauck and Thau 1990:731). This 'acceptability' of the 'vaccine principle' was considered to be 'of particular importance in developing countries' (Jones 1986:184) and is closely connected with another much-cited advantage: the 'ease of delivery within existing health care structures' (HRP 1988:179).

G. P. Talwar of the NII compounds the vaccine image by promoting its immuno-contraceptive with its tetanus carrier for 'additional protection against tetanus, a sizeable health hazard in countries where deliveries do not take place in hospitals but at home in rural areas with high maternal and neonatal deaths' (Talwar 1994a:1). Indian journalist Anjali Mathur points out the disadvantages of the researchers' vision:

> What the pro-vaccine lobby sees as its advantage – ease of administration and acceptability – are really massive disadvantages in India. It is more than likely that harassed family planning staff, under pressure to meet quota set by the government, will give the vaccine to women without informing them about its risk factors properly or properly conducting all the necessary investigations. (Mathur 1993)

Indeed, the 'advantage' of the vaccine image was conspicuously absent from the list of advantages given to women's health advocates at the 1992 HRP meeting with researchers from three of the major research institutions (Griffin 1993:38).[6]

Ironically, the immuno-contraceptive's 'vaccine' image is in some contexts becoming a liability rather than an advantage for the research community. As a Tanzanian physician pointed out at the Medical Women's International Association Conference in May 1995, many people associate a long, if not lifelong, action with vaccines, and would not therefore wish to be 'vaccinated' against their capacity to bear or to beget children.

Any encouragement of an association between an immuno-contraceptive and the positive health benefits of immunization against diseases in fact highlights three interconnected risks: disinformation; potential for abuse; and the difficulties of trusting future immunization, family planning or health care services.

From a user-centred perspective, unbiased and understandable information is an integral part of the provision of contraceptives. Far from being similar, immunological contraceptives are a completely new class of birth control methods which should not be confused with vaccines against diseases.

Low cost

According to the research community, antifertility 'vaccines' are 'likely to be inexpensive when produced in large numbers of doses' (Griffin and Jones 1991:190). Many questions exist, however, about the ultimate costs of the development and introduction of immunological birth control methods.

Manufacturing costs, for example, would vary according to the differing products and formulations and the complexities involved in making them heat-stable in the climates of the countries they are destined for. At the 1989 HRP Symposium on Fertility Regulating Vaccines, it was estimated that the cost of manufacturing one dose of an antifertility 'vaccine' would be comparable to that of modern genetically engineered anti-disease vaccines such as the hepatitis B vaccine. If a new technical method of delivery is employed such as timed-release microspheres, the cost would certainly be higher.

A narrow focus on manufacturing costs, however, leaves out other financial aspects, in particular the *research and development costs*, which at present are largely publicly funded. In addition, research and development itself is behind schedule: HRP had hoped to make its anti-hCG contraceptive commercially available by the mid-1990s (WHO 1986). USAID adviser Jeff Spieler believed in 1987 that vaccines 'interfering with sperm function and fertilization would be available for human testing in the early 1990s' (Spieler 1987:779), whereas they have at most reached the stage of animal trials. Meanwhile, a report from the Indian parliamentary comptroller and auditor-general reprimanded the NII and Indian Institute of Science for their tardiness. The NII was scheduled to finish all three phases of its anti-hCG trials by March 1992, as was Mougdal's team with the male anti-FSH contraceptive (Suresh 1994). An HRP- and Rockefeller-sponsored investigation of how large pharmaceutical companies could be brought back into contraceptive research stated specifically:

> Many company representatives ... believed that contraceptive vaccines would be extremely expensive to develop and, because they are long-acting, would not generate much profit (PATH 1994:416).

The several company representatives interviewed in this study expressed doubt that immuno-contraceptives would be available even within the next two decades (PATH 1994:416).[7]

In any event, a low manufacturing cost does not necessarily translate into a low price for either health services or users. The price charged by a manufacturer will depend on the extent of liability insurance a manufacturer may have to take out (Beale 1991). Liability insurance can be an important cost factor for any biomedical product that carries the risk of adverse effects in a large number of healthy people (that is, a product whose risks are not assessed in relation to a disease). Contraceptives interfering with the immune system may need a relatively high liability coverage compared to other contraceptives.

The market price will also depend on any licensing agreement between the method's developers and the manufacturers. For instance, part of the agreement between the Population Council, the developer of Norplant®, and Leiras, its Finnish manufacturer, is that the implant is sold at one price, approximately $US25, for use in family planning programmes in Third World countries, and at another price, around $US370, for use in countries such as the USA.

The health service providing immuno-contraceptives would have other costs besides that of the products themselves if it were to ensure maximum safety and effectiveness (which still could not be guaranteed). *Service delivery costs* would include those for the several tests needed before and after administering the contraceptive to check for HIV, allergies, auto-immune diseases and a predisposition to these (see Chapter 3), for the frequent tests needed to check the level of antibodies, and for good back-up health care in case of medical problems.

The low price of long-acting methods may not necessarily be an advantage to users because it may result in their preferential promotion without ensuring the necessary service delivery back-up or voluntary and informed choice. Thus users may have to pay a price in terms of health risks and loss of reproductive freedom.

In the case of Norplant®, for instance, the costs of training providers in safe insertion and removal of the implant are a high proportion of the total delivery costs. In subsidized programmes, when some women requested early removal of Norplant®, they were asked to reconsider their decision because of the 'high costs' of the method, while in the USA, women who had had Norplant® inserted for free were delaying its removal even after its five-year period of action because of the $US150 removal charge (Hartmann 1995:212). Removal of Norplant® after five years is essential because the little amount of remaining hormones are insufficient to prevent pregnancy – but enough to increase the risk of a potentially fatal ectopic pregnancy.

Of consideration to the user of a contraceptive is not so much whether it is a low price as whether it is *affordable*. For instance, spread out over one to two years, the total costs to the user of an immuno-contraceptive may well be cheaper than a method that has a shorter duration, but it may well not be affordable if the user has to pay all these costs at the outset.

The low cost of immuno-contraceptives would thus seem not to be certain. Whether or not low manufacturing costs are an advantage for users is a wholly different matter.

A 'population' framework and contraceptive development criteria

Technology is not simply the neutral product of rational technical imperatives; it is the result of a series of specific decisions made by particular groups of people in particular places at particular times for their own purposes ... We can see the effects of social relations that range from fostering or inhibiting particular technologies, to influencing the choice between competing paths of technical development, to affecting the precise design characteristics of particular artifacts ... Technical innovation is profoundly social. As such the technical outcomes depend primarily on the distribution of power in society. (Sociologist Judy Wajcman)[8]

The research conducted during the past two decades has brought us to the threshold of making available a new method for more effectively meeting the challenge of ever-increasing population expansion. (Vernon Stevens, originator of HRP's anti-hCG vaccine, 1986)[9]

From a user-centred perspective, it is doubtful whether the five major 'advantages' cited repeatedly in the scientific literature on immuno-contraceptives are in fact 'advantages'. From such a perspective, there would seem to be little or no advantage in exchanging pharmacological adverse effects for immune-mediated ones, or in having a long-acting method, the user of which cannot know whether and when it is acting and cannot easily stop its action. The supposed acceptability of immuno-contraceptives, a description that centres on the 'vaccine' image, means that it will be difficult for users to trust information and services provided by health professionals. The low costs are anything but certain, particularly if public funding of 'vaccine' and test kit development, liability coverage and service delivery costs are included. From a user-centred perspective, low manufacturing costs are only of value if they translate into the availability of a good product.

However, these five attributes may well appear as advantages from a population control perspective. As Pran Talwar of the NII has said: 'Although a number of methods for fertility regulation are available, their use has not been as widespread as would be desired for effective population control in economically developing countries' (Talwar n.d.:1). When Talwar was interviewed about his motivation and zeal during the Phase II (efficacy) trials, his explanation was simple:

Well, you just have to go, for example, to Bombay, or to any other metropolis for that reason; at the time that the offices close; see this sea of humanity that flows: trains are overloaded, buses are overloaded, everything is overloaded. The population stress is expressing itself in many walks of life.

I would even say that several of our political problems - the uneasiness of the youth, the uncertainties of getting jobs, all the reservation issues – anti- or pro-reservations – are all caused by this problem of too large numbers looking for too few places.

I would even say that the terrorist problem is related in a way to the population problem and the social strain that it is causing - the inability of the structure to cope with the numbers. (in Schaz and Schneider 1991)

In 1988 Australian immunologist A. Basten expressed optimism at the future of this line of research after HRP's first clinical trial in humans:

Fertility-regulating vaccines offer the most practical way of controlling the birth rate, particularly in developing countries. Ideological and political pressure has limited the scope of research being directed towards the goal. Nevertheless, recent advances in reproductive endocrinology, immunology and molecular biology have made an acceptable anti-fertility vaccine a realistic possibility for the next decade. (Basten 1988:771)

Critics of Western medicine have long pointed out that its historical reductionist framework has tended to turn pregnancy into a disease (for example, Martin 1987). A population framework – especially in combination with the 'vaccine' discourse – turns the 'disease' into an epidemic:

Talwar sees population as an epidemic not unlike the tetanus, diphtheria and smallpox epidemics that once ravaged humankind. And it can be defeated, he declares, the same way by a vaccine ... With a vaccine, there would be no remembering to take the pill. No surgery. No struggling with a condom. Pregnancy would be warded off with the same powerful immunological weapons the body uses to fight off disease. (Kanigel 1987:26)

While many and closely spaced pregnancies can be detrimental to women in various ways, the preoccupation with population has resulted in a conceptual framework in which birth control methods are seen not as an entitlement of people but as a weapon in the 'war on population' in which people are treated as mere numbers and statistics to be controlled, manipulated, reduced and dispensed with. In the process, women and men become numbers and statistics and lose any human face.[10]

As the Harvard-educated director of family planning in Bombay, D. N. Pai, declared before joining India's compulsory sterilization programme in 1976:

If some excesses appear, don't blame me ... You must consider it something like a war. There could be a certain amount of misfiring out of enthusiasm. There has been pressure to show results. Whether you like it or not, there will be a few dead people. (quoted in Hartmann 1995:243–4)

In contraceptive research and provision, a preoccupation with 'over-population' and population growth rates of certain groups of people, particularly poorer people, nonwhites and those from Third World countries, has been present for most of this century. Executive Health Officer Kathuria, of India's Bombay-based Population Project 5, is clear as to who the targets of birth control methods are:

... this class of people – especially in the slums – who have four and five children. The well educated ones are not having more than one child; and they are absolutely in line with the developed places. But it is these people, who are in the slums and in the lower economic group, they are spoiling the demographic pattern of Bombay and India. (in Schaz and Schneider 1991)

Outright coercion to use contraception or be sterilized is now said by many in population organizations or those working in family planning programmes to be a thing of the past, but use of military metaphors is not. The WHO's magazine *World Health* describes antifertility 'vaccines' as '*attractive* additions to the world's family planning armamentarium' (Griffin 1987:25, emphasis added). Scientist Rodney P. Shearman, in a foreword to a book by Warren Jones, describes them as 'antigenic weapons' to disrupt the 'reproductive process, a process that left unchecked threatens to swamp the world' (in Jones 1982:vii). John Aitken, who is working on anti-sperm and anti-egg contraceptives, believes that 'new methods of birth control are necessary to halt, and ultimately reverse this inexorable trend in population growth' (Henderson et al. 1987).

It is not surprising, therefore, that a perception of immuno-contraceptives as 'vaccines' against uncontrollable pregnancy 'epidemics' and as weapons in the 'family planning armamentarium' results in a framework of contraceptive assessment and development in which consideration of the freedom and well-being of individuals is less of a priority. Yet HRP's former director, Mahmoud Fatallah, actually promoted HRP's immuno-contraceptive at the 1992 HRP meeting between scientists and women's health advocates by likening it to a 'smart bomb': it would affect reproduction and nothing else. (I wondered whether one day their adverse effects be classified as 'collateral damage'.)[11]

At the HRP 1989 symposium N. Avrion Mitchison acknowledged the participants' diverse views on immuno-contraception in his chairman's summary:

> Some see it as little more than an extension of the existing methods of medium-term contraception, such as the Norplant®; others feel that the potential for ease of administration, coupled with cheapness and long-term effect, place it in an entirely new category. Some regard it as tampering with germ cells and the complexities of the immune system, and therefore extremely dangerous; others see it as safer than steroids, because it immediately stops something from happening, rather than continuously blocking a normal activity. Some hope for a vaccine simply because it will provide women with a worthwhile new birth control option, while others hope that it will provide an unprecedented effective instrument of demographic control. (Mitchison 1991:249)

Mitchison himself, however, although recognizing that 'an antifertility vaccine would need to be equivalent, in terms of risk–benefit, to established family planning methods, and will need to pass safety tests of utmost stringency' (1991:248), was preoccupied with another matter:

Foremost in my mind during these discussions was our difficulty in assessing the urgency of the demographic crisis. To the extent that the impact of that crisis increases, the need for more effective family planning technologies must increase. At the very least, failure to develop something that may provide a more effective technology would be to take a grave and unnecessary risk. (1991:250)

As a participant in the 1989 symposium, the notion that new contraceptives are needed in order to diffuse the 'population crisis' seemed curious to me. Why did women in developing countries have to be 'persuaded' to use birth control when throughout the ages and in many cultures women have wanted and found the means to have a greater influence over their fertility?

Prior to the 1989 symposium I had been working as a lecturer in community pharmacy in a Thai university and then as a public health researcher with a Bangkok-based consumer protection group. I was aware, therefore, about distortions in the range of pharmaceutical drugs developed and in the information provided to 'consumers' because of the profit motives of the pharmaceutical industry. I had participated in a worldwide campaign for the wide availability of essential drugs and the provision of unbiased information.[12]

In 1989 I knew little about the influence of population ideology on the development and provision of birth control methods, or, conversely, about the reinforcement of the notion of 'effective population control' by the availability of modern contraceptives. As the senior adviser to the Population Council, Sheldon Segal, a major player in the development of Norplant®, puts it: 'It is perhaps no coincidence that official family planning services began around 1960, the year that modern IUDs and the Pill were introduced for the first time' (Segal 1993:11).

I subsequently learnt that population ideology – intertwined with a narrow biomedical view of human reproduction – has had a major influence on contraceptive research, shaping the trend towards a long-acting 'model' method, a result of which would be that 'unwanted pregnancies owing to failure of patients to use the method properly would be greatly reduced' (Stevens 1990:345).[13] The outcome is that women have had fewer and fewer opportunities to prevent outside control of their fertility. I learned that high statistical effectiveness had taken priority over safety.

I learned to see how the focus on birth control methods as the major tool of demographically oriented family planning programmes had led to a framework of thought in which people had become 'targets' or, at most, passive 'acceptors' of contraceptives, whether they wanted them or not.

Meeting women's needs – or forestalling their resistance?

Much has happened in the contraceptive research and population control institutions since the 1989 HRP Symposium on Fertility Regulating Vaccines. In the run-up to the 1994 UN International Conference on Population and Development, important key figures started to call for a profound reorientation of contraceptive research. Mahmoud Fatallah, Senior Adviser for Biomedical Health Research at the Rockefeller Foundation and former director of the World Health Organization's Human Reproduction Programme (HRP), for example, declared in a widely circulated paper:

> The first contraceptive technology revolution was demographic driven with emphasis on the development of methods that were effective, long-acting, and widely available ... For the second contraceptive technology revolution, the field must again be goal driven, and the goal should be set right, focusing sharply on a sustained effort to develop contraceptives that address the still unmet needs of women. The demographic impact will not be diminished, but enhanced. The message for all concerned about population growth would be clear: women know best. (Fatallah 1994:229)

Fatallah called on the research community to reconsider seriously research priorities, given the financial constraints in the field which, he said, may call on researchers 'to drop leads' and to refocus on urgent requirements that are poorly met by existing contraceptives (Fatallah 1994:230).

A shift away from demographic goals towards 'meeting women's needs' and 'increasing their choices' was advocated as a guideline not only for decision making in contraceptive research. The director of population sciences at the Rockefeller Foundation, Steven Sinding, advised family planning policy makers on the occasion of the fortieth anniversary of the International Planned Parenthood Federation 'to replace [demographic] program targets with objectives expressed in terms of stated desires of the people served' (Sinding 1994:27). If one believes the official report of the 1994 UN International Conference on Population and Development, the current era is now one of 'population policies based on reproductive rights'.

Such statements can readily give the impression that policy makers have finally taken on board the concerns of all those critics who have been pointing out that population thinking has greatly harmed people by fostering the belief that none of the world's major problems can be solved without 'population policies'.[14]

Social justice activists are sceptical, however, about this supposed shift on the part of policy makers and institutions towards 'meeting women's needs'. As participants at a 1993 meeting on People's Perspectives on 'Population' in Bangladesh stressed:

Women's basic needs of food, education, health, work and political participation, a life free of violence and oppression, should be addressed on their
own merit. Meeting women's needs should be delinked from population
policies, including those expressed as apparent humanitarian concerns for
women. Women should have access to safe contraception and legal abortion
under broader health care. (Declaration of People's Perspectives on 'Population' Symposium, 12–15 December 1993, Comilla, Bangladesh)

Neither Fatallah nor Sinding ever called for a delinking of 'meeting
women's needs' from 'population control'. On the contrary, Fatallah
maintained that 'respecting women and responding to their needs is one
of the best strategies for saving the planet. The demographic impact
will not be diminished, but enhanced' (Fatallah 1994:229). Such a view
accords with Sinding's claim that current analysis 'strongly suggests
that … helping women and couples achieve their desired family size
would accomplish as much, or more, demographically than meeting most
country targets' (Sinding 1994:27).

In case some family planning policy makers are still reluctant to
abandon setting or meeting demographic targets, Sinding points out
that population targets carry the seeds of coercion and that the coercive
policies 'produced severe backlashes in a few places as citizens' organizations reacted angrily to efforts to compel adoption of particular
methods of family planning. A reaction against government targets for
adoption of contraceptives and quotas for field workers emerged'
(Sinding 1994:27). But as many concerned people have pointed out,
meeting women's needs so as to meet (even unspoken and unspecified)
demographic targets is, at most, a shift from 'hard' to 'soft' population
control (Hartmann 1995:154).[15]

The trend towards 'softer' population policies has been going on for
more than ten years now. In terms of contraceptive service delivery, it
has been translated into a call for a shift of focus from the 'hardware'
approach – the development of newer long-acting contraceptives –
towards a 'software' approach: increasing the 'quality of care' of the
services provided. This shift was an initial recognition that people are
not simply passive 'failing' users of contraceptives, that many people
avoided family planning services either because of their experiences with
the side effects of the contraceptives they had been provided with or
because of careless or disrespectful provision of contraceptives. This is
not to discount the work of all those concerned health professionals to
ensure a better quality of care. But in practice, a shift from focusing on
the 'hardware' of family planning to its 'software' without addressing
the impact of population ideology has meant primarily that it does not
matter what type of birth control method people use – so long as they
use something.

This shift should have had some implications for a population-centred assessment of the benefits of antifertility 'vaccines'. In both 'hard' and 'soft' frameworks, worry would be expressed about the 'risk' of not developing a technology to tackle population growth, seen in terms of a 'crisis', and any birth control method would primarily be assessed as a tool to reduce birth rates.

In a 'hardware' approach, it is not surprising that developers – such as the researchers of the Population Council's Phase I (safety) anti-hCG product – came to see 'ease of administration, lack of user failure, long-lasting protection and low cost' (Brache et al. 1992:1) as 'major' advantages of this birth control technology. In a population-control-centred framework, antifertility 'vaccines' seem indeed to come close to the ideal method of mass fertility control.

In a 'software' approach, however, population planners may be concerned about their introduction. Given the reactions of people to immunization programmes, such as those described at the end of Chapter 4, and an acknowledgement that abuse leads to resistance, policy-makers may prefer not to introduce a method that may increase people's distrust of family planning services.

If a major concern is, in contrast, the well-being of future users, an assessment of risks may have a completely different approach. Graham N. Dukes, a leading expert on the risks and safety of medicines, sees the development of immunological birth control methods as 'asking unnecessarily for trouble'. He says that 'whatever risks there are can hardly be predicted in any test. But what we know of physiology suggests that they could be very serious' (Dukes 1995).

A women–centred contraceptive – for population control?

From a user-centred perspective, it is a major problem that contraceptive development has not been oriented towards women's reproductive self-determination. This does not mean that individuals engaged in contraceptive research and family planning intend to harm people or to disregard their health and well-being. Nor does it mean that all those engaged in contraceptive research and family planning wholeheartedly embrace population thinking. The trends of thinking in contraceptive research and family planning are as diverse as in women's movements. As in the early days of family planning (see Introduction), mainstream family planning organizations include many people who regard contraceptives primarily as a means of decreasing maternal mortality and some whose major focus is women's self-determination – but they also include those who are eager to curtail the birth rates of particular classes or 'races' of people.

Many of those engaged in contraceptive research may never have reflected on the framework of thought shaping their decisions, while others may not see a conflict between developing contraceptives to control population and developing contraceptives to meet women's needs.

It seems time to ask whether it is possible to design a contraceptive which could meet both the goal of population control *and* that of women's reproductive self-determination. In my opinion, a framework of contraceptive development for which the focus is a reduction of birth rates yields very different criteria for assessing the 'benefits' and 'risks' of contraceptives compared to a framework centred on enhancing women's reproductive rights and well-being.

The major 'benefit' that population control methods aim for is to facilitate easy and effective mass fertility control. In this framework, the reliability of a method from an individual's point of view is not considered, adverse effects are assessed primarily in terms of their impact on the 'continuation rates' of users, and contraceptive abuse becomes a problem only when people stop using contraceptives.

A women-centred assessment, on the other hand, is intended to provide contraceptives that enhance women's space for decision making, that interfere as little as possible, if at all, with complex body systems, that carry little risk of unwanted long-term infertility or sterility, and that cannot be used by outsiders to control women's fertility against their wishes. A women-centred framework also asks for far more efforts to be exerted towards developing male contraceptive methods in order to enable men to share the burden of birth control.

A reorientation of contraceptive development and provision towards a focus on women's empowerment and reproductive self-determination cannot be achieved simply by adding a few more contraceptives to the existing gamut. A 'contraceptive revolution' must go beyond increasing contraceptive choices; the challenge is to *improve* the range of available contraceptives. This involves a re-evaluation of all birth control methods including those that are now widely distributed and those that are still in the pipeline.

Notes

1. In Griffin and Jones 1991:190.
2. There is no denying that menstrual disturbances can have significant effects on women's social and sexual lives and any absence of these would be welcome. Menstrual disturbances caused by contraceptives are often classified as 'minor' adverse effects. For instance, a report evaluating Norplant®'s safety states that 'while many women report irregular bleeding, these occurrences are not medically harmful' (Townsend 1990). Such assertions are based on measures of immediate blood loss and haemoglobin levels, not on studies of the long-term effects of continuous or frequent disruption of a woman's menstrual cycle, including effects on her health and well-being. Moreover, in cultures where a

woman is considered 'unclean' during her bleeding days, she may not be allowed to take part in traditional ceremonies or prayers or various other aspects of daily life or to have sexual intercourse (UBINIG 1990; Hanhart 1993). In contrast, while some researchers may think that amenorrhoea (no periods) would not be a problem for women and might indeed be welcome, such a belief ignores the fact that menstruation and menstrual patterns are also indicators of health and ill-health and may be important for a woman's sense of identity. This belief also ignores the time-honoured fact that a lack of menstruation is regarded by many women as an indicator of pregnancy rather than its reverse.

3. A report by the Committee on Contraceptive Development, drawn from several US institutions, on the obstacles and opportunities in developing new contraceptives contains a typical example of the purported benefits of new methods in which the user appears as little more than a number or statistic. The reason for lessening the adverse effects would seem to be to increase the number of users and the length of time they are contracepting for: 'New methods can influence fertility in several ways – by increasing safety (yielding fewer adverse effects) and effectiveness (yielding fewer pregnancies), increasing acceptability and use (yielding more users), and increasing continuation (producing longer durations of use). Because most modern methods of contraception are relatively effective, the impact of new methods will probably come from greater acceptance, longer periods of use, or both' (Mastroianni et al. 1990:27). The report goes on to calculate and compare the 'demographic impact' of increased continuation of use and increased number of people who use contraceptives ('contraceptive prevalence'), concluding that 'If continuation rates were 90 percent during the first years of use instead of the 50–70 percent for most temporary methods, it would make a substantial difference' (Mastroianni et al. 1990:27).

4. When the Pill was first introduced, two of its oft-cited advantages were its convenience and its facilitation of 'spontaneous intercourse'. These are now increasingly being cited as advantages of long-acting methods. However, women have found that the Pill did not so much allow *them* a more spontaneous and enjoyable sexuality, but rather was taken by men as a sign of their sexual availability, irrespective of whether the women themselves wanted sexual intercourse or not. Some women have pointed out that methods used only during coitus, such as the condom or diaphragm, have increased their sense of self-worth, and have opened up possibilities for discussion with their partners. Today, an increasing number of women and men choose barrier methods – even if they do not consider them the most 'convenient' methods – because they are the best option in terms of safety, the condom having an additional advantage of protection from STDs and HIV infection.

5. Barriklow 1993:29.

6. According to David Griffin's briefing paper at the 1992 meeting, 'Sufficient information was already available at that time [i.e. when research into immuno-contraceptives restarted] to indicate that it should be feasible to develop fertility regulating vaccines that:

— 'are free of overt pharmacological activity and the metabolic, endocrine and other physical disturbances that often accompany other methods of birth control;

— can confer mid- to long-term (3 months to 1–2 years) but not permanent protection following a single administration;
— are easy to administer without manipulation of the genitalia;
— remain effective without continuous conscious action by the user;
— are inexpensive.' See Griffin 1993:38.

7. Calculation of the ultimate research and development costs would also have to include the costs of developing diagnostic tests needed to determine the effectiveness of each particular immuno-contraceptive as well as the costs of research into the safe reversal of each formulation.

8. Wajcman 1994:154.

9. Stevens 1986:374.

10. See the challenging analysis by Duden 1992.

11. Thanks to Mila Abramovicz for pointing out this parallel of making human victims of technologies 'disappear'.

12. For an excellent account of these issues, see Chetley 1990.

13. It is not uncommon in the literature on contraceptive research for completely healthy people receiving contraceptives in 'clinical trials' to be referred to as 'patients'.

14. There is insufficient space in this publication to provide either a detailed review of the history of population control institutions and of people's experiences of population programmes, or a demonstration of how population thinking has distorted the analysis of deeply political problems such as poverty, the exploitation of peoples and nations, and environmental degradation, and contributed to a loss of human dignity and basic rights for millions of people. Putting any problem into a population framework automatically leads to the making of associations between the problem and demographic dynamics – even where there is none. Moreover, such action distracts attention from questions such as why the model of economic growth is still being pushed all over the world, despite the knowledge that it is a major cause of the depletion of natural resources. For further information, see Furedi, forthcoming; Hartmann 1995.

15. Betsy Hartmann writes of the new rhetoric surrounding population policies, '... in the absence of real social transformation, the emphasis will probably be on the "motivational efforts" to sell the idea of small families. Many of these messages will push the consumer model: with fewer children, you can buy more and degrade the environment less, which of course is a doubtful proposition. Social marketing of contraceptives, rather than the establishment of comprehensive health services, will continue to be a priority' (Hartmann 1995:154).

It is doubtful to what extent the change in rhetoric surrounding population policies will be matched by a change in the content of such policies. In 1994, in the Indian state of Andhra Pradesh, for instance, village revenue officials tried to prevent people from harvesting their crops by placing flags in the fields of those people who refused to be sterilized (*Political Environments* 1995:30).

6

A question of ethics

The leading principles to be honoured in medical and social sciences are to obtain informed consent from the research subjects and to minimize the risks to which the research subjects are exposed. Furthermore, most rules and guidelines also express a positive evaluation of new knowledge; when human research subjects are involved, only experiments promising some new knowledge should be performed at all. (Stellan Wellin, director of the Gothenburg Centre for Research Ethics, 1993)[1]

Critics have long claimed that the value and belief systems underlying contraceptive development have adversely affected not only the direction of contraceptive research but also the research practice (see e.g. Corea 1991). The research community, however, maintains that whatever their interests, they are bound to respect internationally recognized standards of ethics in human trials. Any violation tends to be explained as an 'isolated case'.

The question of ethics is less an issue of personal motivation on the part of researchers than of the responsibility of the research community towards trial participants and future users of their products, a responsibility that includes making efforts to detect any biases shaping research practice. How does research on antifertility 'vaccines' fare if one assesses it in accordance with some of the major criteria of ethics in biomedical research, in particular the criteria of the World Medical Association's Declaration of Helsinki?

Do the advantages justify the risks?

The 1964 Declaration of Helsinki (last revised in 1989) states in its preamble:

The purpose of biomedical research involving human subjects must be to *improve* diagnostic, therapeutic and prophylactic procedures. (in CIOMS 1993:47, emphasis added)

Basic principles 4 and 5 specify further that 'biomedical research cannot legitimately be carried out unless the importance of the objective is in proportion to the inherent risk to the subject' and that ultimately:

> Every biomedical research project involving humans should be preceded by careful assessment of the predictable risks in comparison with foreseeable benefits to the subjects and others. Concern for the interest of the subject must always prevail over the interests of science and society. (CIOMS 1993:48)

Consideration of these principles raises the question of whether immuno-contraceptives will present advantages over existing contraceptives or not, that is, whether their anticipated benefits justify the risks not only to participants in the trials but also – and above all – to the major projected users of the method.

From the perspective of the individual user, it is actually difficult to find any benefits of immuno-contraceptives, considering that alternative methods are available. The principle of action of immuno-contraceptives – their reliance on the human immune system – leads to an efficacy profile that, as we have seen, is particularly unsuited for individual contraception, having the following drawbacks:

— an initial lag period of at least around two weeks before the method is effective as a contraceptive;
— probably a relatively low method effectiveness;
— unreliability because of unpredictable variations in immune responses. It is doubtful whether in real life conditions, test kits (if they can be developed) would solve this problem;
— the difficulty, if not impossibility, of safely 'switching off' immune reactions.

With such an efficacy profile, it is difficult to see how exposure of women and men to the potential adverse effects of immunological birth control methods such as auto-immune diseases, allergies, immune-complex diseases and interaction with pre-existing diseases, and exposure of foetuses to ongoing immune reactions, can be justified.

The long-lasting action of an immuno-contraceptive, combined with the impossibility of immediately reversing it and the potential of its easy mass administration without the knowledge or consent of the 'users' form the basis of an unprecedented potential for immuno-contraceptives to be abused – whether or not this was the intention on the part of those designing and researching the method. Indeed, the potential for abuse alone seems sufficient reason to question the continued development of antifertility 'vaccines'.

The probable consequences of these contraceptives' high abuse potential include an individual's loss of self-determination over his or her fertility as well as deprivation of bodily integrity and dignity. They also include: health consequences for women, men and children because of a probable tendency for service providers to skip tests before administration; a shift of 'choices' of contraceptive offered by service providers towards long-acting methods with biased information; a loss of trust in the health services; and an increased perception of pregnancies as 'epidemics' and of women as spreading these epidemics.

Finally, as we have seen, immuno-contraceptives offer no benefits in terms of protection against sexually transmitted diseases, but may in fact increase the risk of HIV transmission through unsterile needles (if the contraceptive is in the form of an injection) and interfere with public health programmes to encourage condom use.

As such, the decision to pursue the development of immunological birth control methods would seem to contradict major tenets of ethics in biomedical research.

As most researchers seem to have made a different assessment as to the anticipated benefits and risks, a further question is to what degree the Helsinki Declaration was followed in the trials themselves. Although most of my study has concentrated on a prospective technology assessment of immuno-contraceptives from a user-centred perspective and not on the research practice, there are several indicators to suggest that at least some of the research is not conforming to these internationally accepted standards. Nonconformity would seem to be particularly apparent in the areas of minimizing risks, animal trials, scientific principles and informed consent. What follows are just a few examples, rather than a comprehensive review, of the research practice.

Minimization of risks

[T]he seriousness of any of the potential hazards, and the recognition that a fertility regulating vaccine is an entirely new contraceptive principle, dictate extreme caution in proceeding to clinical trials with any such vaccine. (Warren Jones, principal investigator of HRP Phase I trial)[2]

Risks to trial participants (as well as of course to any final potential users) could have been minimized in several areas, in particular: selection of the target antigen; the means used to increase the effectiveness of an immuno-contraceptive; the criteria used for evaluating contraceptive efficacy; and the prevention of abuse.

Selection of candidate antigens   As indicated in Chapter 2, a major factor influencing the risk–benefit balance of specific immuno-contraceptives is the selection of the reproduction antigen to be used. In 1977, early on in the research, HRP's Task Force on Immunological Methods for Fertility Regulation, in consultation with specialists in immunology, toxicology, and reproductive biology and representatives of three national drug regulatory authorities, drew up guidelines for testing the efficacy and safety of model contraceptives. These stressed that:

It is essential, since they are to be used by healthy people, that all methods of fertility planning be thoroughly tested from the point of view of safety. In the case of an antifertility vaccine, which represents a totally new area of intervention and whose effects are intentionally of long duration, this is even more important. This poses a major problem, since guidelines for safety evaluation do not exist. (HRP 1978:360)

Four principles for the development of antifertility vaccines were specified:

— to be effective, the target antigen should be essential for the reproductive process;
— to minimize the risk of auto-immune disease, the target antigen should be restricted to the intended target (i.e. hormone, egg cell, sperm cell);
— to minimize the risk of immune-complex disease, the target antigen should not be present continuously in the person who receives the 'vaccine', but only intermittently and/or in low concentrations;
— the antifertility effect of immunization should not be permanent and there should be no demonstrable hazard to offspring born subsequently to vaccine users. (Ada and Griffin 1991a:xvi)

The first three principles centred on the definition of the 'perfect' target antigen, whose neutralization through immunological means should result in an effective antifertility response without creating immune-mediated disturbances.

Twelve years later, prior to Phase II (efficacy) trials of its latest anti-hCG immune contraceptive, HRP called a second meeting, held in June 1989 at the WHO headquarters in Geneva. Besides reviewing the principles of antigen selection and discussing further trial phases, the intention in calling the meeting was to discuss 'social, legal and ethical issues which the preparation and use of these vaccines would raise so that appropriate action could be taken to avoid potential problems in this area' (Ada and Griffin 1991a:xvii).

To this meeting HRP invited lawyers, social scientists and consumer representatives, representatives of all five major research institutions, as well as immunologists, reproductive biologists and representatives of national drug regulatory authorities.

At this meeting, the third of the principles previously established for immune contraceptive research stating that the target antigen should not be present continuously in the person who receives the 'vaccine' was changed. The amended principle stated that '*Preferably*, the molecules should be present transiently and in relatively low amounts so as not to overwhelm the predicted immune [antibody] response' (Report

1991:254, emphasis added). In addition, the principle was reclassified as an efficacy criterion whereas previously it had been a safety criterion.

The 1977 principle would have excluded from development all immuno-contraceptive methods targeting non-pregnancy-associated hormones (GnRH and FSH) which are present continuously in the body, and those anti-hCG prototypes that might cross-react with LH. By altering the principle, any exclusion was abolished, despite concerns expressed by a number of participants at the 1989 symposium.

Reproductive biologists T. Chard and J. S. Howell from St Bartholomew's Hospital Medical College in London affirmed that 'on the basis of the animal evidence ... all of the hormones of the hypothalamic–pituitary–gonadal axis [i.e. GnRH, FSH and LH] would be excluded'. For them, the hCG-CTP being researched by HRP was 'the only significant candidate for an anti-fertility vaccine for the foreseeable future' (Chard and Howell 1991:111). Participants at the 1989 seminar stressed in their final report that 'the unknown consequences of chronic immunity to "self" molecules in the brain, pituitary and gonads would argue against using immunogens restricted to these sites' (Report 1991:255).

But just a few years later, human trials of immuno-contraceptives which use these previously unacceptable target antigens were underway. The Population Council and the NII are now developing anti-GnRH products for men, although in 1983 some researchers had questioned why such a method was being developed given that chemical substances that block GnRH are available and 'can be more precisely controlled and ... easily reversed' (Anderson and Alexander 1983:560).

So far the US Food and Drug Administration (FDA) has not permitted the Population Council to embark on clinical trials on an anti-GnRH formula as a contraceptive, but only as a treatment for prostate cancer. In this case, the aim is to prevent the production of testosterone which can be a contributing factor to the growth of prostate cancer. The risks of the product under investigation are thus weighed against its benefits in treating the cancer.

However, the interruption of testosterone production should be seen as an undesired adverse effect in the case of an immuno-contraceptive. Although researchers promise to develop a testosterone slow-release injection or implant to 'replace' the hormone, can such a combined approach truly justify the risks of immunological neutralization of GnRH? The many unresolved question of adverse effects aside, who would ensure the availability of testosterone replacement to men in poor countries?

In India, NII has tested an anti-GnRH contraceptive on at least twenty women who had just given birth 'to prolong post-partum amenorrhoea' (Talwar 1992:513). Fellow researchers at the 1992 HRP meeting openly condemned this research because infants could be exposed to anti-GnRH antibodies through breast milk. Under international ethics guidelines, research with infants and nursing women is allowed only under very restricted circumstances.

Pran Talwar later stated that the trial batch of the anti-GnRH contraceptive had been stored inappropriately and had therefore been inactive at the time of the trial, but that he planned to repeat the trials in nonlactating women (HRP 1993:27). The quality of the research is in doubt if the product was badly stored, while the trial itself cannot be justified according to ethical standards because anti-GnRH immunocontraceptives to prolong lactational amenorrhoea will never have an advantage over the oldest and healthiest contraceptive in the world: breastfeeding. As the research community knows, 'breast feeding [on demand] provides more than 98% protection from pregnancy in the first six months' (HRP 1990:65) (see also Table 3.1).

At the 1992 HRP meeting, some scientists agreed with women's health advocates that immunological anti-GnRH contraception was inappropriate, not only because of the need for constant replacement of testosterone when used by men and oestrogen when used by women, but also because of the risk of potential immune-mediated damage in the brain and pituitary gland (HRP 1993:22). For some researchers, however, anti-GnRH products for male contraception offer a particular promise – that of early access to industry funding, as they could also be sold as anti-cancer drugs (Thau 1993).

The NII has since exposed 148 women to its particular anti-hCG (oLH) immuno-contraceptive, despite the fact that the US FDA advised against Phase II (efficacy) trials with the doubly cross-reacting antigen (Hari 1994:96).[3]

Meanwhile N. R. Moudgal of the Indian Institute of Sciences has tested anti-FSH immuno-contraceptives in men, although the benefits of this method are more than questionable because FSH is no longer considered essential for sperm production (HRP 1990:30).

One reason for this 'diversification' of research into immunocontraceptives with different target antigens is that in the United States the Population Council ran afoul of the anti-abortion movement with its proposed development of an anti-hCG immuno-contraceptive. During the Reagan and Bush presidential eras, the New Right managed to stall all US government funding for any research or service that could be related to abortion, be it family planning services or technology development. Anti-hCG 'vaccines' were seen as abortifacients, because hCG is produced only after conception.

As John Aitken, working on anti-egg and anti-sperm immune contraceptives with the Reproductive Biology Unit of Edinburgh University, has said, an anti-hCG contraceptive 'disrupts pregnancy, effectively working as abortion, and to many it will be unacceptable'. He maintains that 'a more constructive way is to block fertilization and prevent pregnancy from occurring at all' (Hope 1992:28).

Like Aitken, other research teams are now concentrating their research on anti-egg and anti-sperm contraceptives, although their development is judged to be the most difficult. Avrion Mitchison, however, encouraged researchers at a 1989 CONRAD international workshop on anti-egg and anti-sperm contraceptives as follows:

> I want to re-emphasize that finding good candidates [antigens] is relatively cheap compared with the enormous cost of developing and testing a vaccine. This argues against going for broke at too early a stage. But, on the other hand, funding is competitive, and the earlier we have something to show for our efforts, the more likely we are to secure further support. In this sense a prototype vaccine is needed, even though we know that it may not be the optimal choice and may never enter widespread use. (Mitchison 1990b:612)

Credit has to be given to HRP for attempting to set standards for this internationally dispersed research enterprise. Unfortunately, these few examples would seem to indicate that minimization of risk does not seem to have been uppermost in the minds of most of the developers when selecting the target antigen.

Means used to increase effectiveness Besides the selection of the target antigen, the formulation of immuno-contraceptives – such as the choice of, for example, carrier, adjuvant and 'vehicle' – has implications for their anticipated risks. For example, the use of live viruses such as an altered vaccinia virus or salmonella virus as a means of making the antigen appear foreign to the immune system is controversial, even for anti-disease vaccines, not least because of potential transmission to other people and species (McNally 1994:209). According to Talwar, '[the live recombinant] vaccines would be highly cost effective and suitable for the Southern countries' (Talwar 1994a:2). Such viral vectors also harbour the 'potential drawback' of 'the possibility of inducing infections and/ or permanent sterility' (Stevens 1992:139).

As shown previously, different research teams have opted for different approaches to tackle a fundamental problem of anti-hCG contraceptive research, namely the problem of achieving a satisfactory contraceptive efficacy (see pp. 24–7, 33–4). Immunologist Avrion Mitchison expressed his concern about the known cross-reactions between the NII's anti-hCG product and LH and about the risks of the 'drastic' adjuvants used in HRP's formula – but concluded: 'that concern diminishes as the number of women who have been vaccinated without adverse consequences increases' (Mitchison 1990:726).

Such a statement seems to represent a fundamental shift in the basic logic of clinical trials, which holds that the onus is on researchers to ascertain as far as possible that the apparent or anticipated benefits will outweigh any known or anticipated risks *before* human trials are carried out – not during them or after.

Criteria for contraceptive efficacy   The major focus of the efficacy and safety discussion at the 1989 symposium centred on the criteria for antigen selection, an obvious focus if the technology is considered as a vaccine (i.e. an immunological method for mass administration) rather than as a contraceptive to be used by individuals. Throughout the research, the setting of standards for the minimization of risks and maximization of the products' benefits as contraceptives seem to have received less attention.

The other consumer advocate invited to the 1989 HRP symposium besides myself, Anita Hardon from WEMOS, a Dutch group involved with issues of women and pharmaceuticals, expressed her concern:

The proposed Phase II and III vaccine trials in India and the trials promoted by the WHO Special Programme are premature. Clinical trials should only be conducted if the new methods offer significant advantages over already existing methods. Neither vaccine yet meets this criterion, and it remains open whether they will do so in the future. (Hardon 1989)

HRP's David Griffin responded that Phase II (efficacy) trials in fertile women were important to show the feasibility of the whole approach:

If the ongoing and planned clinical trials with anti-hCG vaccines are success- ful, a family planning method which is effective for a period of approximately 18 months as the result of a single injection and which does not produce any menstrual cycle disturbances or other side-effects (apart from the transient soreness at the injection site common to all intramuscular injections) would be a realistic proposition. We feel that such a method would be an attractive addition to the currently available technologies and would offer a number of advantages to both the users and providers of family planning services. (Griffin 1990b:7)

Yet statements made about the ultimate efficacy characteristics of immuno-contraceptives researchers are aiming for, in particular in terms of their duration, the number of injections, and effectiveness in prevent- ing pregnancy, are contradictory and ever-shifting.

Sometimes the duration is cited as 18 months (Griffin 1990b:7) and sometimes as '3 months to 1 to 2 years' (HRP 1993:38). Researchers refer to a 'single injection' (HRP 1993:11, 38), 'one or two injections' (HRP 1988:179) or a 'course of immunization' (Ada and Griffin 1991c:18).

When some researchers refer to a one-year action, it is often not made clear that this is the duration they are aiming for with anti-hCG methods. The duration for other immuno-contraceptives may be different. For example, John Aitken, from the Reproductive Biology Unit at Edinburgh University, projects that his anti-egg contraceptive may last for around four years (Hope 1992) and his anti-sperm immuno-contraceptive for women three to four years (Cookson 1991).

It cannot be guaranteed, however, that frequent exposure through coitus will not boost the action of anti-sperm 'vaccines' for life. 'That's where I fall off the bandwagon,' said Bill Brenmer, director of the Center for Population Research at the University of Washington in Seattle recently. 'I am not convinced that any [anti-sperm] vaccine can be used as a reversible contraceptive' (Vines 1994:40). Brenmer's doubts are shared by Nancy Alexander, Chief of the Contraceptive Development Branch of the National Institute for Child Health and Development (NICHD), Deborah Anderson of Harvard Medical School, and Faye Schrater, who is now a steering committee member of HRP's Task Force on Fertility Regulating Vaccines (Anderson and Alexander 1983: 567; Schrater 1994a:259).

But to be acceptable as a contraceptive, an immuno-contraceptive would obviously need to be effective in preventing pregnancy. Griffin and Jones have stated that:

> To be an attractive method of fertility regulation, the annual pregnancy rate associated with the use of the vaccine should not exceed 2 percent, similar to the most effective reversible methods already in use. However, for a preliminary demonstration of the efficacy of this approach, an annual pregnancy rate of at most 5 per cent would be acceptable and encourage further development of this line of research. (Griffin and Jones 1991:188)

Despite the figures cited, this is actually rather a vague statement about the method's effectiveness. In 1983 Deborah J. Anderson and Nancy Alexander pointed out the difficulties of ever reaching a response in 95 per cent of the immunized persons with an immuno-contraceptive based on a single antigen (Anderson and Alexander 1983:567).

As with the duration of an immuno-contraceptive, different researchers seem to have their own ideas of what is sufficient method effectiveness. Even the goal of a 2 per cent method failure rate does not appear to be a binding commitment, either. A 1995 position paper on HRP's programme cited as a design criterion 'effective contraception for 6 months in at least 95% of users' (Gevas 1995:1) while Australian immunologist A. Basten, who works closely with HRP, has given a yardstick of a 'minimum efficacy rate of 90%' (Basten 1988:760). Avrion Mitchison comments that Talwar and his team have demonstrated that immuno-contraception can be 'remarkably effective ... but only in women who make the right immune response' (Mitchison 1993:108). The women who did not 'make the right response' were some 20 per cent in the Phase II trial.

It is not just so as to be 'an *attractive* addition' that immuno-contraceptives 'would have to be as effective as the best of currently available alternative methods' (Griffin 1993:39, emphasis added). The overall effectiveness, the number of injections for primary immunizations and the duration of action are not just discretionary values. A bottom line for a minimum method effectiveness is particularly important for any 'systemic' birth control method – contraceptives that interfere either with our hormones or with our immune systems – because of their potential effects on a foetus.

As outlined in Chapters 2 and 3, the number of primary immunizations and duration of action has implications in terms of the risks of allergies, immune-complex diseases and the probability of exposure of the foetus (as well as potential for abuse). Any shift in the criteria of efficacy has different implications in terms of the risks for the user. The model duration of one year harbours the risk of long-term adverse effects – and a high potential for abuse. The longer the duration of action, the higher these risks. While a shorter action may decrease the risk of abuse, it increases the probability of foetal exposure and the probability of allergic reactions and immune-complex diseases. Many of these problems are compounded by the difficulty of 'switching off' immune reactions. Experts who had been asked to prepare a briefing paper for the 1989 HRP symposium on future needs in 'vaccine' development stressed that 'Increasing emphasis will need to be placed on achieving prolonged (i.e. for 1–2 years) immune responses following a single administration of the vaccine' (Basten et al. 1991:89). The imprecision over efficacy criteria raises three questions: what is really feasible? Where is the point at which an immuno-contraceptive is considered a failure? In fact, is there any acceptable duration of action?

Reversibility The fourth of the principles for the development of anti-fertility vaccines stated in 1977 of HRP's efficacy and safety guidelines, relating to the nonpermanence of immunological contraception, was reaffirmed at the 1989 HRP symposium.

But before HRP's Phase I (safety) trial, its principal investigator, Warren Jones, wrote this:

> The capability for reversal is an attractive but not essential facet of any contraceptive method and its abrogation by a vaccine would at worst only limit the utility of such an approach. (1982:16)

According to Jones, 'Early experience with an hCG vaccine … suggests that the practical duration of immunity following primary immunization with this antigen is 12–18 months.' Yet in the case of anti-egg and anti-sperm immuno-contraceptives, 'the ultimate place of a contraceptive vaccine may indeed evolve as a form of medical sterilization; ... the prospect of permanent infertility might be circumvented by gamete [semen or egg] storage'. Jones did concede that 'this principle ... is logistically unattractive as an adjunct to a widely employed contraceptive method' (Jones 1982:16).

HRP's Task Force Manager David Griffin has stated that 'vaccines with a shorter duration of effect may prove attractive alternatives to injectable contraceptive steroids [that is, hormonal injectables], and vaccines with a longer duration of effect may be attractive alternatives to surgical sterilization' (Griffin 1990a:508). For Rosemarie Thau of the Population Council, 'Irreversibility ... is not always an adverse effect; some vaccines may be designed to be used as non-surgical means of sterilization' (in Mauck and Thau 1990:728). If the research community concedes the possibility that some immunological birth control methods may ultimately evolve into immuno-sterilizants, there must be a totally different medical and political debate.

Prevention of abuse   When the 1977 efficacy and safety guidelines for research into (anti-hCG) immuno-contraceptives were set out, any consideration of potential abuse of the method was absent. At the 1989 HRP symposium, however, the working group on the 'social, ethical and regulatory aspects of antifertility vaccines' stressed that the 'potential risks and benefits of antifertility vaccines need to be assessed ... from the viewpoint of potential misuse and abuse' (Report 1991:290). This was a landmark recommendation – but one that has been hidden away by limiting discussion during the working group itself and in published reports of the meeting.

The HRP-appointed chair of the particular working group, John Dunne,[4] said that 'clearly the vaccine will be abused' as he went on to curtail debate among participants by cutting short controversies and admonishing participants to use 'positive' language so as not to put the pharmaceutical industry off the product.[5]

The draft report of this 1989 HRP meeting explicitly recommended 'that upon registration [for general use], a detailed free and informed consent procedure outlining the particular features of the new method be developed, appropriately updated and used consistently' and that the 'development of the consent procedure should involve user representatives', an accurate summary of the working group's recommendation.

Yet in the edited, final published report, possibly because the editors had not understood the recommendation, it was placed between two paragraphs dealing with clinical trials; as a result, it could easily be construed as being a recommendation for clinical trials only, and not as originally intended for use after market approval of immuno-contraceptives (Report 1991:294). Placed into a section about trials, the recommendation loses much of its ground-breaking value because the need for free and 'voluntary informed consent' is already a recognized mainstay of any guideline on ethics in clinical trials.

Had the recommendation been placed in the section dealing with 'abuse risks' (Report 1991:292), its meaning, as intended by those who made the recommendation, would have been much clearer. Family planning programmes do not generally require their users to sign informed consent forms before receiving a contraceptive; such a recommendation therefore gives some indication of concern about the unknown long-term health risks and the higher potential for abuse of immuno-contraceptives.[6]

Instead, the final published report of the 1989 HRP symposium asserts that the 'abuse risks associated with antifertility vaccines are *the same* as those that apply to all methods of birth control' (Report 1991:292, emphasis added) – a statement that was not made in the report of the relevant workshop group, and that is actually contrary to the opinion expressed by several symposium participants, including the two consumer representatives, Anita Hardon and myself.

The only proposal published in the report to 'prevent administration of the contraceptive without the consent or knowledge of the recipient' was as follows:

> In order to keep them distinct from other vaccines, such as those used in mass vaccination programmes against infectious diseases, it is important that antifertility vaccines are distributed through family planning programmes and services. However in some countries family planning services and vaccination services are provided by the same clinics and often by the same individuals. Under these circumstances it may prove difficult to ensure that antifertility vaccines are kept distinct from other vaccines and there is a risk that they may be administered incorrectly, as one of a group of antidisease vaccines. (Report 1991:292)

Although the stated aim of the 1989 HRP meeting was 'to avoid potential problems' in the social, legal and ethical arenas, given the disparity between what was said in this particular workshop and what was published in the final report, one might question whether the aim was in fact to avoid potential problems not for users but for future immuno-contraceptive development and future 'acceptability' of the method.

Animal trials

> Biomedical research involving human subjects must conform to generally
> accepted scientific principles and should be based on adequately performed
> laboratory and animal experimentation and a thorough knowledge of scientific
> literature. (Helsinki Declaration, Basic Principle 1 [CIOMS 1993:48])

In the development of any drug, animal studies precede the human
trials in order to minimize the risks to human beings. Such studies are
relevant only if the animal reacts in a similar way to how a human being
can be expected to react. Standards for ethical research thus maintain
that the 'animal model' must be appropriate.

Immuno-contraceptives pose yet another challenge in this respect.
For testing the anti-hCG immuno-contraceptives, primates are the only
animals sufficiently closely related to human beings that produce chorionic gonadotropins (CG). Ideally, the requisite animal tests should be
carried out in 'higher' primates such as chimpanzees or gorillas. But as
these species are protected because of their dwindling numbers worldwide, the most frequent test animals are baboons or rhesus monkeys.
But the baboon immune system reacts far less to an immuno-
contraceptive based on human CG than to one based on species-specific
baboon CG.

For years, HRP stressed that a prerequisite for its Phase II (efficacy)
trials was the development of an immuno-contraceptive for baboons
based on the baboon pregnancy hormone (bCG) instead of the human
one (hCG). The team considered that only this anti-bCG immuno-
contraceptive trial was valid to 'assess the risks to the foetus should an
unexpected pregnancy occur' (HRP 1990:27).[7]

But a baboon analogue has since been quietly dropped. 'You shouldn't
think we haven't tried,' said David Griffin to women's health advocates
at a 1992 HRP meeting, adding that HRP had spent more than
$US200,000 attempting to develop such an anti-bCG contraceptive.
However, Vernon Stevens, the originator of HRP's anti-hCG immuno-
contraceptive, said at the meeting that such a trial was still needed
(personal communication; HRP 1993:26).

HRP and the Swedish investigators of the Phase II trials thus decided
to proceed without testing the effect of anti-baboon CG antibodies on
baboon foetuses. They did opt, however, for a second-best option which
was to test baboons with the product based on human CG for its
teratological effects (potential to harm a foetus). But when the product
was being prepared for such studies, a 'significant proportion' of the
freeze-dried product would not dissolve in the oily emulsion. HRP then
decided to postpone the teratology studies in baboons until after the
Phase II (efficacy) trials in humans, with the agreement of the Swedish
and US drug regulatory authorities (HRP 1991:99–100).

In the trial itself, the brochure that was intended to inform trial participants about the product being tested contains this statement:

> No adverse effects of anti-hCG antibodies on the foetus could be demonstrated in animal tests, but it is not known what the effects will be in humans. In order to participate in these trials a woman must therefore be prepared to terminate her pregnancy if this contraceptive fails. In such a case the aborted material will be examined for the presence of antibodies and potential damage to the foetus. If, however, the woman decides to continue her pregnancy we will monitor her very carefully. (Anti-hCG-vaccin 1993:7)

One might question whether it can be called a 'free and informal' decision, if women are pressured to abort and at he same time reassured about adverse effects on foetal development when the animal trials referred to in the brochure were not designed to check for potential adverse effects on the foetus. There have been specific teratology tests in rats and rabbits (HRP 1992:92). However, it was pointed out at the 1989 HRP symposium that 'the use of standard tests in small animals (such as rats and rabbits) is irrelevant for testing for possible immuno-teratological effects of a vaccine' (Jones and Beale 1991:148).

Back in 1976, there had been controversy over an early human trial of an immuno-contraceptive conducted in India by the NII. It was described as unethical because it had exposed unsterilized women to a product that had not been appropriately tested for potential effects on the foetus. Indian scientist K. S. Jayaraman reported in the journal *Nature*:

> Although Talwar was the first to put the hCG vaccine into human trials in 1974, he lost the race because of controversies that cropped up after he jumped the gun. In a hurry to beat his competitors, he vaccinated six unsterilized women with hCG-TT [tetanus toxoid] vaccine when its efficacy was still in doubt. Two of the women became pregnant, the World Health Organization withdrew support and questions on ethics raised by the Indian scientific community forced him to go back to the laboratory and animal trials. (Jayaraman 1986:661)

Talwar complained that HRP's own interests had prompted it to withhold further funds for his research as part of a bid by Western scientists to undercut research by Third World researchers (Tripathi 1979).[8]

The question as to whether animal trials can yield relevant predictions about the effects of a product in humans seems now to have become less of an issue among developers of immuno-contraceptives. Some researchers explained to women's health advocates at the 1992 HRP meeting that 'differences would still exist between the baboon and human situations and that the increasing volume of data being generated in clinical trials is more relevant than animal data' (HRP 1993:26).

That same year, French researcher Dominique Bellet, who is developing his own anti-hCG contraceptive at the Gustave-Roussy Institute independently of the major research teams, announced plans to skip the test phase in monkeys altogether, passing directly from rats and rabbits to women, 'because of the great safety' of his synthetic anti-hCG formula (Joras 1992:8; Vigy 1992).[9]

Although immune reactions and reproductive processes vary enormously between different species and one may question whether there is an appropriate animal model at all for the anti-hCG immunocontraceptives, it seems as if some researchers may be taking the requirements of animal tests somewhat too lightly, drifting towards a tendency to replace animal trials with human ones.

Conformity with scientific principles

Trials on any novel products should follow principles that allow their biomedical characteristics to be assessed correctly. In their background document on clinical trials presented at the 1989 HRP meeting, Jones and Beale advocated variations in the design of Phase I clinical trials so as 'to accommodate unavoidable geographic and cultural logistic difficulties and to avoid an unacceptable imposition on the health status of volunteers'. For instance, they referred to the relatively high number of blood samples taken during clinical trials in Australia and concluded 'this would be untenable in anaemic Indian women' (Jones and Beale 1989). (The published version of this presentation states that 'such extensive sampling schemes might be inappropriate in other populations' [Jones and Beale 1991:155].)

By definition, Phase I (safety) trials should be limited to healthy women. Pran Talwar has stated that 90 per cent of the women in India are anaemic and that their immune status is far below that of their Western counterparts (Talwar 1989, personal communication). It is unclear whether or not Talwar limited his Phase I trials to healthy women.

Informed consent

In any research on human beings, each potential subject must be adequately informed of the aims, methods, anticipated benefits and potential hazards of the study ... The physician should then obtain the subject's freely given consent, preferably in writing. (Helsinki Declaration, Basic Principle 9 [CIOMS 1993:49])

The protection of the health, well-being and rights of trial participants is at the very core of ethical guidelines for human trials. The investigators have a duty to ascertain that any person participating in a trial of a new biomedical product does so in the full knowledge of a method's anticipated benefits and risks, and without being put under duress.

Enrolment of some women in the Phase II NII trial in India of anti-hCG contraceptives in 1990 was documented by German film-maker Ulrike Schaz. The film shows one woman being told:

> We have got a new injection ... the effect of the injection stops children for one year ... You need not be afraid about this. The injection has no side effects. You see this injection is absolutely 100% effective ... We'll also put in a copper-T [IUD]. Continuous copper-T is not very good. If you have it three years, six years, then there is the risk of cancer. That's why we want you to change. (Schaz and Schneider 1991).

(The IUD was inserted to cover the lag phase and was removed after three months.) The women are then asked to sign a consent form written in English, although 'only a few of the women can understand and read English' (Schaz and Schneider 1991).

Schaz was told by a doctor at the trial that this type of enrolment was not exceptional and that she herself felt she was insufficiently informed about the risks and benefits of the immuno-contraceptive being tested.

For its Phase II (efficacy) trial, however, HRP together with the Swedish research teams it was working with designed a comprehensive information brochure in nonmedical language for a potential 250 participants. Yet this brochure still has serious shortcomings according to international standards of informed consent. It does not stress sufficiently the novelty of immuno-contraceptives, which are compared to anti-disease vaccines which prevent 'illness ... and even death'.

Antifertility 'vaccines' are described as an 'attractive method of family planning' because they may have less adverse effects than hormonal methods or IUDs, but may be more effective than barrier methods such as condoms or diaphragms. However, to prove the 'theoretical advantages over some of the currently available methods' ('teoretisk flera fürdelar framfür dagens preventivmethoder'), the brochure makes a highly selective comparison: to mention only the adverse effects of hormonal methods and IUDs but not the lack of adverse effects of barrier methods, and to highlight the theoretical effectiveness of immuno-contraceptives only in comparison with that of barrier methods (which many people presume to be less effective) and not with that of hormonal methods and IUDs is to provide biased information. In order to decide whether or not to participate in the testing of a new drug, a woman should be able to make an informed decision as to whether the supposed benefits of the product being developed are worth taking the accompanying risks for.

The selection in question, however, gives the impression that there is no need to worry about major risks and that immuno-contraceptives will be relatively effective. A truly informative comparison of the 'advantages' of immunological birth control methods should have involved a comparison with both their benefits *and* their risks.

Trial participants were not informed about the omission of the baboon teratology trial (see p. 104).

The brochure states that women 'can discontinue participation' at any time ('Anti–hCG-vaccin' 1993:8). Indeed, one of the principles of informed consent is that the trial participant 'is free to withdraw his or her consent to participation at any time' (Declaration of Helsinki, Basic Principle 9). But in one respect, this is simply not possible with immuno-contraceptives because of the very nature of the method's action. A woman may decide to stop participating in the trial, but any immune-mediated contraceptive and adverse effects could continue in her body irrespective of her participation or consent, possibly for a considerable time. For meaningful informed consent, women about to participate in a trial should be specifically informed of this.

Given all the potential risks and uncertain benefits of immuno-contraceptives, so long as human trials continue any women (and men) participating in them should, as a minimum, be informed about the experimental nature of immunological methods for birth control, their unreliability as contraceptives and the uncertainty surrounding their safety and reversibility.[10] Of course, this may lead to a distinct shortage of 'volunteers'.

Conclusion

> Above all, it must be remembered that, unlike therapeutic drugs that are used to treat specific diseases, family planning methods are intended for use by healthy individuals for prolonged periods of their reproductive life. This imparts an added dimension of ethical responsibility to the conduct of clinical trials of all new methods of fertility regulation including antifertility vaccines. (Report of working group on clinical trials at the 1989 HRP Symposium on Vaccines for Fertility Regulation [Ada and Griffin 1991a:280])

Considering the novelty of turning the immune system against human components for contraceptive purposes, this brief glimpse into some of the research practices reveals alarming tendencies. The health and safety criteria for target antigens seem to change according to technical feasibility and availability of funding, whilst efficacy and abuse have received far less attention, animal trials have been skipped, and women have been enrolled into trials on the basis of biased information.

Notes

1. Wellin 1993:68.
2. Jones 1982:17.
3. The FDA plays a particular role in international contraceptive research. Most research institutes apply for Investigative New Drug Status with the FDA, even if both the research institute and place in which the trials are to take place are outside the USA. Because this regulatory authority is well-respected, approval of clinical trial protocols by the FDA can facilitate later approval of the product, not only in the USA but also in other countries.
4. John Dunne is a member of the WHO Secretariat Committee on Research Involving Human Subjects, and Chief of WHO's Pharmaceutical Unit.
5. A major part of the discussions of this working group centred on issues of liability during trials and after the product had been marketed.
6. An exception is IUD use in the USA, although the issue in this case is liability for risks rather than prevention of abuse. An informed consent form is not, of course, a failsafe guarantee against coercive or unwitting administration, but it might go some way towards deterring too casual an administration of immuno-contraceptive methods. Such a document could also provide women with evidence that the immuno-contraceptive had been administered, if they ever wanted to take out liability suits against the contraceptive manufacturer or provider. This was a problem for many women who experienced severe difficulties with the Dalkon Shield IUD: when the manufacturer agreed to pay compensation, they had no proof that it had actually been inserted in them. In drafting such a consent form for participants in immuno-contraceptive trials, care would have to be taken that the form was not written in such a way as to absolve the manufacturer or provider of any responsibility if any problems developed.
7. The animal model was a major point of discussion at the 1989 HRP symposium. See Griffin and Hendrickx 1991:32.
8. Today however, Talwar is at the forefront of antifertility 'vaccine' research and participates in steering committee meetings of HRP's Task Force on Vaccines for Fertility Regulation (HRP 1992:146) as well as those of the Population Council.
9. To my knowledge, he was not able to do so because of protests expressed through the press.
10. For other conditions, see Report 1991:272–3.

7

Fostering a public debate

> We must not let decisions in matters that intimately affect us be determined by experts whose impartiality is far from clear. Of course, we must be able to choose whether or not to use new technologies. But more important, we must become part of the processes that determine what technologies should be developed. (Ruth Hubbard and Elijah Wald 1993:127)

> A feminist approach to birth control would have to take into account the present range of options women have, the context in which those choices are made and what future developments we want to see ... After all we must insist that birth control, like other health care, exists for us, not we for it. (Sue O' Sullivan 1987:353)

By June 1995, a worldwide coalition of over 430 groups and organizations from thirty-nine countries was demanding that research and development of antifertility 'vaccines' be stopped immediately. This chapter summarizes the evolution of an international campaign to encourage public debate on whether antifertility 'vaccines' should be developed or not and the responses of the research and funding communities to the campaign's demands.

The beginnings of unease

Several individuals and groups have been concerned for some time now about the development of immuno-contraceptives.[1] Women's movements, in particular, have been apprehensive about this line of research. In 1987, Brazilian feminists initiated the collection of 10,000 signatures of concerned citizens, including 300 scientists, against the Population Council's proposed testing of its anti-hCG 'vaccine' in Brazil. The testing was subsequently abandoned (Gomes dos Reis 1988:12; 1994).

Participants at a 1990 International Women's Health Meeting in Manila, and at a 1991 Rio de Janeiro Meeting of the Feminist International Network of Resistance to Reproductive and Genetic Engineering (FINRRAGE) passed declarations that this line of contraceptive research should be stopped (Philippines Organizing Committee 1991:151; FINRRAGE 1991).

Other women's health advocates pressed for more information from the researchers about what was being developed. For example, women's health activists invited in February 1991 by HRP and the International Women's Health Coalition to map out 'Common Ground' between women's advocates and contraceptive researchers asked HRP to convene a special meeting to discuss the 'development of contraceptive vaccines' (HRP and IWHC 1991:41).[2] In April 1991, many of the participants from over twenty countries at a conference exploring women's perspectives on contraceptive development and use (organized by the Women and Pharmaceuticals Project of a Dutch organization, WEMOS) questioned what antifertility 'vaccines' would actually mean in practice for women. During a conference workshop on the politics of contraceptive development, a suggestion was made that a report should be produced on antifertility 'vaccine' development to generate public discussion on the issue.

The BUKO Pharma-Kampagne (a German group which campaigns against global malpractices of multinational pharmaceutical companies and for rational drug use and policies), its Swiss counterpart within another organization, the Berne Declaration, and the European co-ordinating office of Health Action International raised funds to produce German- and English-language versions of the report.[3]

Researcher's dream: woman's nightmare?

Just over two years later, in June 1993, the BUKO Pharma-Kampagne hosted a two-day conference in Bielefeld to launch the report *Vaccination against Pregnancy* (Richter 1993) (which this book is based upon). The report concluded that 'unless concerns about safety, effectiveness, predictability, reversibility and potential for abuse can be met, this research should not be continued. Unless a clear advantage can be shown over existing methods, human trials cannot be justified' (Richter 1993:65). Some hundred participants attended the conference to explore whether 'vaccinations' against pregnancy were a researcher's dream but a woman's nightmare.

Presentations and plenary discussions centred on the appropriateness of immunological contraceptives given the prevalence of AIDS and their anticipated impact on women's reproductive self-determination. Zambian immunologist and general secretary of the Society for Women and AIDS in Africa Nkanda Luo said she saw no place for immunological birth control methods when HIV infection rates in some places were as high as 25 per cent of the adult population and when many people in Africa were suffering from chronic infections. The most pressing issue, she said, was not to introduce new birth control technologies but to reorient family planning services, in particular, to promote the use of condoms and responsible male sexual behaviour (*Pharma-Brief* 1993:3).

Kalpana Mehta from the Indian women's rights group Saheli had no doubt at all that many Indian researchers and policy makers would consider an immuno-contraceptive effective in only 85 per cent of users to be a 'success' because they tended to regard poor women as statistical figures who caused the birth rate to increase rather than as human beings. She criticized the trend in India of continuously introducing more long-acting contraceptives in family planning services oriented towards population control while the general health care system in the country was being dismantled because of structural adjustment programmes imposed by the International Monetary Fund and the World Bank (*Pharma-Brief* 1993:3).

The explanations of HRP's David Griffin as to why HRP regards antifertility 'vaccines' as a positive expansion of women's choices stood in marked contrast to the concerns of participants.

Participants concluded the conference by spontaneously passing a resolution calling for a stop to the development of antifertility 'vaccines' (*Pharma-Brief* 1993:5).[4] Immediately after the conference, nineteen women activists from twelve countries retreated into the countryside for an Action Workshop to explore how to translate their concern into concerted action in as democratic a way as possible.[5, 6]

Outlining a campaign framework

All the workshop participants insisted that one priority was to stop research on immuno-contraceptives as soon as possible because of the potential effects on women enrolled in clinical trials and the apparent lack of information being given to trial participants about the novelty of the method. All the activists were worried that the method's immense potential for abuse would mean that once available on the market, antifertility 'vaccines' would be beyond any form of regulation or control.

Much thought went into the general framework of an international but decentralized campaign that would respect the autonomy and specific national contexts that campaigners have to operate in. Particular attention was paid to the issue of how to demand that this particular line of contraceptive research be stopped without such a demand being interpreted as being against all forms of contraception or against abortion. This issue was of particular concern for women from Latin American and Caribbean countries who – in contrast to many women in Asia who have been inundated with contraceptives in population control programmes over the last three decades – are still struggling to obtain more access to contraceptives because of the political influence of, among others, the Catholic Church. The group decided that, while objecting to antifertility 'vaccines', they would always stress that they were in favour of the development of safe and user-controlled contraceptives to be delivered in a system guaranteeing voluntary and informed use.

The framework in which contraceptive research is carried out was also discussed. Participants from India explained how tired they were of having to campaign each time a contraceptive was introduced in the country that they felt was not primarily oriented at women's health and rights but at reducing population growth. Many felt that, with many pressing issues to deal with, the introduction of the newer contraceptives diverted too much of the already overextended energy of women's groups – irrespective of whether they tried to prevent a method being introduced or whether they cooperated with the institutions to ensure a method's best possible introduction and provision. Workshop participants concurred that historical trends in contraceptive research and service delivery made it necessary to stress the political context in which anti-fertility 'vaccines' have been developed.

Other participants emphasized the need to question the reductionist scientific framework that seems to guide the work of many contraceptive researchers. They felt that a mechanistic view of the human body was as significant in explaining current trends in contraceptive research as was the availability of funding within a population framework. Only the quest for the 'ideal' population control method, combined with the dynamics of modern scientific enterprise, could explain the large investment in antifertility 'vaccine' research and in other long-acting birth control methods and the little attention paid to the improvement of barrier methods, the development of male fertility control methods and investigations into 'natural' and traditional birth control methods and practices.[7]

All these reflections were contained in a petition drafted by participants, intended to be used as a tool to raise public awareness and to pressure the research and funding community (see Appendix 3 for the complete text and Appendix 4 for a list of signatories). The petition – a Call for a Stop of Research on Antifertility 'Vaccines' (Immunological Contraceptives) – expresses concern about the method's abuse potential, questions the legitimacy of the manipulation of the immune system for contraceptive purposes given the distinct lack of advantages but significant potential risks of the method, and draws attention to specific problems in the clinical trials that had taken place so far. It criticizes the shaping of contraceptive research by population ideology and a narrow scientific framework and concludes with a dual demand, namely, an immediate stop to all research on immunological contraceptives and a radical reorientation of contraceptive research.

A campaign is launched

Participants at the Bielefeld Action Workshop would have liked to consult with more activists before circulating the petition and launching the campaign to raise public awareness. Yet when the pace is set from outside, in this case by the research institutions, there tends to be an unfortunate tension between optimum consultation and effective action. Out of concern for women participating in clinical trials, the group decided to send the petition to the major institutions researching and funding immuno-contraceptives as soon as possible and to publicize it through the mass media at the same time, unless those attending two further workshops at the International Women and Health Meeting in Uganda in September 1993 recommended otherwise (which they did not). Beatrijs Stemerding of the Women's Global Network for Reproductive Rights (WGNRR) agreed to become the first campaign coordinator.

With less than three months to go from the winding up of the initial international consultation process to the date chosen for the start of the campaign, 8 November 1993, some of us considered postponing the launch because only a handful of signatures had endorsed the petition. But a few days before the launch, signatures started pouring in. The fax machine at the office of the Women's Global Network hardly stopped. The Call for a Stop had been circulated at several regional and international conferences such as the International Women and Health Meeting in Uganda and the Latin American Feminist Encuentro in early November 1993 in San Salvador. By 8 November, 232 groups and organizations from eighteen countries had signed the petition.

A wide variety of activities around the world marked the start of the campaign. The WGNRR, in cooperation with WHAF, organized an international press conference, whilst many other organizations held local press conferences, as a result of which media in many countries took up the issue (see, for instance, Levene 1993).

In India, sixteen women's and human rights groups staged a demonstration in front of the Asian Regional WHO office in New Delhi with a banner bearing the words 'You can't treat pregnancy like a disease!' They presented the petition and a list of its signatories to representatives of the WHO's regional office, the Science Ministry's Department of Biotechnology and the National Institute of Immunology. At first, Pran Talwar, the principal 'vaccine' researcher at the NII, was warm and cordial. He explained to the delegation that the petition was a smear campaign against him by Western feminists and the WHO, which was jealous of his success. When the Indian women explained that they were working with these 'Western feminists' and had themselves been involved in drafting the petition, Talwar quickly curtailed the conversation, promising to prove the quality of his 'vaccine' research to the women at a later meeting (Mehta 1993).

The Bombay-based Forum for Women's Health organized meetings for women's groups and the press to give out information about the 'vaccine'. Other Bombay groups organized protests at public meetings, declaring 'No jobs, no education, no food, no health – but contraceptives, contraceptives, contraceptives!' (Mathur 1993). They appealed to the general public:

> The campaign against antifertility vaccines is ... not merely a women's issue, as issues of contraceptives have so far been seen. For us, it is an issue of human relations, of the whole understanding of what is development and the meaning of progress in science and technology. (Forum for Women's Health 1993)

As already outlined, groups from Latin America and Africa faced the difficult task of calling for a stop to research on immuno-contraceptives when women have difficulties obtaining modern contraception and safe, legal abortion. Yet groups such as the Brazilian SOS Corpo, the Colectivo Mujer y Salud (Women and Health Collective) in the Dominican Republic and the Women's Health Group of the South African Health and Social Services Organization (SAHSSO) organized a range of activities to publicize the issue. Other groups, for example in Canada, Germany and the Netherlands, called on their governments to stop funding this line of research.

The next major event at which the campaign was publicized was the 1994 World Health Assembly, the annual meeting of the World Health Organization's policy-making body, held in May 1994 in Geneva. On the opening day of the assembly, the Swiss groups Antigena and Espace Femmes International staged a street theatre performance in front of the conference building. More than fifty women from all over Switzerland put together five human tableaux depicting the realities of contraception: 'situation' portrayed a group of pregnant women; 'propaganda' played out a population-control-oriented family planning programme; 'research' was illustrated by a scientist behind a gigantic microscope and a doctor injecting two women; 'protest' indicated a women's demonstration; and 'resistance' was exemplified by two women breaking a huge syringe. Large banners called on the WHO to stop research into antifertility 'vaccines'.

Judging by the reactions of assembly delegates and journalists covering the meeting, it was clear that few of them had ever heard of the antifertility 'vaccine'. Yet many felt it was a topic deserving wider discussion. The action was reported in more than thirty articles in the Swiss press alone. Meanwhile Health Action International-Europe and WHAF distributed a flyer 'Immunological Contraceptives: Designed for Populations, not for People' in the conference halls (HAI/WHAF 1994).

The research and funding community reacts

By early 1995, most of the research institutions and a few funders had responded in writing to the Call for a Stop. Some researchers also gave direct or indirect reactions in journal articles. Some common characteristics of these reactions can be identified, namely, the assertions that there is no need to be concerned about the abuse potential and that any problems can be ironed out as the 'vaccines' are improved; the various attempts to deligitimize critics; and the claim that the 'vaccines' are being researched so as to improve women's choices.

Claim 1: no need to worry over abuse potential   The petition calling for a stop to research into immuno-contraceptive research maintains that:

> Immunological contraceptives will not give women greater control over their fertility but rather less. Immunological contraceptives have a higher abuse potential than any existing method'. (Appendix 3)

No other section of the petition has received more reactions than this one. Giving an initial reaction to the petition to the press, Population Council spokesperson Sandra Waldman said that 'it has been proven by independent research that the more methods are available, the more people will use them, simply because women and men need options' (in Levene 1993). As indicated in Chapter 5, however, increasing the number of users *per se* is not an advantage to users – but may be to those concerned to maintain high rates of contraceptive prevalence.

In its official position paper on the 'vaccines', the Population Council is more reflective:

> Any provider-dependent contraceptive method could potentially be used in an abusive fashion ... Such potential could be publicly debated and safeguards taken in any country where any new contraceptive technology is introduced. Furthermore, the use of any new technology should be carefully monitored over time in order to ensure that systems of informed consent are in place and are functioning effectively. Since the vaccines that are currently being tested have a duration of effectiveness of less than one year, they are less likely to be open to abuse than sterilization or other long-acting contraceptives. The system of informed consent used worldwide for sterilization is a good model to follow if and when any vaccines reach the introduction stage. (Population Council 1994:1)

At first glance, such statements seem as though the Population Council has taken on board some of the concerns of activists concerning the actual use of contraceptives and the circumstances in which they are provided. For instance, it does concede that 'provider-dependent' contraceptives can be abused and takes this possibility seriously enough to suggest informed consent forms should be used, which is not a usual practice in contraceptive provision.

However, the proposed safeguards against abuse do not respond to the point that some contraceptives are more prone to abuse than others; moreover, the extent to which international informed consent has been effective in preventing abuse of sterilization itself, particularly in the case of vulnerable persons, is doubtful.

By suggesting that abuse and prevention of abuse are matters solely of contraceptive provision and programme implementation, the Population Council lays the responsibility for potential abuse on to others, ignoring its own role in developing a product with technology-inherent features that lend it to abuse. Absolving itself in this way while apparently acknowledging abuse allows the institution to continue its research.

Instead of focusing on abuse potential, Talwar argues against the assertion that immuno-contraceptives will not give women greater control over their fertility by citing the catch-all 'miracle' word – 'choice':

> At present the vaccine formulations are of three types. One produces an effect for only three months, the other for six months and the third (under test in monkeys at present) has sustained antibody titres for one or two years. The user will have the choice of any of the formulations for the period desired. The birth control method is not provider-dependent. As with insulin for diabetics, the woman can inject herself. (in Sunny and Shah 1994:27)

Both Talwar and the Population Council confuse the issue of control and abuse by referring to current antifertility 'vaccines' (anti-hCG types) that are effective as contraceptives for less than one year. This is less than the intended duration of action that is being aimed for.

Even in its report of the Phase II (efficacy) trial, the NII team implicitly disclaims any technology-inherent risk of abuse of antifertility 'vaccines' when it concludes that:

> The present study demonstrates the feasibility of a birth control vaccine that women can choose on a voluntary basis ... Since antibody titres decline spontaneously unless booster injections are given, the duration of effective immunization can be controlled by the woman herself. (Talwar et al. 1994a:8,385).

Such a statement blurs the distinction between reversal at will and reversibility after the contraceptive phase.

The director of the US-based CONRAD, Henry Gabelnick, and the HRP team reacted more strongly to the campaigners' assertion that antifertility 'vaccines' will have an unprecedented potential for abuse. Gabelnick wrote:

It strikes me that the assumption of ill intent inherent in this fear is inappropriate and largely unjustified ... By assuming the worst about people's behaviour and designing our lives so that protection from the ill intent of others is the guiding principle, we establish a fortress mentality and a paranoid society. (Gabelnick 1994)

As pointed out in Chapters 4 and 5, concerns about mass abuse – illinformed, unwitting or coercive administration – of birth control methods do not necessarily imply that those developing the method are intentionally planning such abuse. Given that technologies can shape society as well as society can shape technologies, public debate on any envisaged new technology should anticipate the potential outcomes of its development, including the issue of social accountability for adverse outcomes.

Given the present struggles of US citizens to prevent the enactment of legislation compelling certain categories of women to have Norplant® inserted and the obstacles many women face to even engage in such struggles, one can only respond to Gabelnick in the words of Nicolien Wieringa from the Women's Health Action Foundation:

For many women, at this very moment, contraceptive abuse is a reality in their lives. Fighting this situation needs different approaches, both by ending the structures that enable people to abuse contraceptives and by preventing the marketing of contraceptives that can potentially be abused. (Wieringa 1995)

HRP's David Griffin and his colleagues in effect refute the suggestion that different birth control technologies can have different potentials for abuse depending on their various characteristics. They maintain that 'all existing and new methods are potentially open to abuse ... The risk and incidence would not be reduced by stopping the development of antifertility vaccines' (Griffin, Jones and Stevens 1994:108). While it is no doubt true that contraceptive abuse may continue irrespective of whether 'vaccines' are developed or not, the assertion does not respond to the critics' claim that abuse may actually be increased if the vaccines do become available.

Griffin, moreover, turns the argument of abuse against critics themselves by claiming that 'it's just as abusive to prevent women from having access to preparations they want as it is to give them preparations they don't want' (in Levene 1993). The HRP team subsequently stated that 'Restricting contraceptive choice to the currently available methods and thereby forcing people into using methods they would prefer not to use, or worse still, not use any method at all, would be another form of abuse' (Griffin, Jones and Stevens 1994:113).

Leaving aside the question of whether the research institutions investigated women's and men's wishes when research into antifertility 'vaccines' started up, statements such as these blur the distinction between the abuse of birth control methods and the lack of (or denial of access to) contraceptive services. Any support of people-centred access to services has to assess whether some birth control methods actually decrease rather than increase the potential for reproductive decision making. Reproductive rights groups have long campaigned simultaneously for user-centred development and provision of birth control methods and against unsafer and abuse-prone contraceptives.

Thus, none of the responses from the research and funding community adequately refutes critics' assertions that the antifertility 'vaccine' will lead to women having less control over their fertility, not more, and that the method will be more prone to abuse than other methods; instead, the reactions confuse the issues or deny responsibility.

Claim 2: antifertility 'vaccines' can still be improved   Overall, the institutions do not accept that there are any grounds for concern about the ultimate biomedical characteristics and risks of immuno-contraceptives; they maintain that as they are still under development, they can be improved.

HRP's current director Giuseppe Benagiano, for example, asserts that in the animal and Phase 1 (safety) human trials of HRP's anti-hCG prototype, 'a high level of efficacy was obtained, no menstrual cycle disturbances or other adverse side effects were observed, and the vaccine was well accepted by the clinical trial volunteers' (Benagiano 1994:1). This assertion depends on how 'a high level of efficacy' is defined and is in direct contradiction to a report on the Phase 1 trial that mentioned abnormal menstrual patterns in 4 of the 20 women injected with the prototype agent (Jones et al. 1988:1297). In addition, the 'principal adverse reaction' in the women – muscle and joint pains – did not cause 'any of the volunteers to withdraw from the study' because some of them decided to take painkillers (HRP 1988:182).[8]

Benagiano goes on to accuse the authors of the Call for a Stop of 'alarmist speculation, technical and scientific inaccuracies, and distortions of facts to support their opposition to the development of new technology' (Benagiano 1994:1). His letter does not give any indication of what these inaccuracies are or proof of such, but refers instead to an article HRP was invited to write by the editorial board of the journal *Reproductive Health Matters*. The article, which was published five months later (Griffin, Jones and Stevens 1994), in fact does not give any indication of inaccuracies in the critics' descriptions of the current characteristics of immuno-contraceptives, but instead disagrees that these characteristics pose insoluble problems.[9]

The HRP authors state that the current immuno-contraceptives are only 'prototype preparations' and that 'another five to ten years of further development and careful pre-clinical and clinical testing will be needed to determine if anti-hCG vaccines can be considered for introduction into family planning programmes' (Griffin, Jones and Stevens 1994:110). They maintain that HRP has 'established the feasibility of developing anti-fertility vaccines' and that the product merely has to be optimized (Griffin, Jones and Stevens 1994:113).

At issue, however, is not whether immuno-contraceptives can be improved but whether they can ever be improved sufficiently to be both safe and effective in the real life conditions of future users and their potential children.

CONRAD's Gabelnick declared that 'we will not abandon research on immunocontraception unless scientific evidence demonstrates that the methods are not safe and effective or if it becomes apparent that there are not advantages over existing long-acting technology' (Gabelnick 1994:1).

Do we really have to wait until human trials or use after approval by drug regulatory authorities provide irrefutable 'evidence' of the unreliability or adverse effects of immuno-contraceptives? As indicated throughout this book, the current state of knowledge about the immune system, vaccines against diseases and women's varying experiences with family planning methods would indicate that it is highly unlikely that immunological birth control methods will offer any advantage to the user that could justify their development.

Annette Will from the BUKO Pharma-Kampagne points out that insisting on scientific evidence curtails the wider dimensions of what is at stake:

> We are debating the introduction of a new technology, and we consider it crucial not to focus the debate on laboratory research only and neglect experiences in other fields as well as the intentions of research. (Will 1995)

Women's health advocates attending the 1992 HRP meeting wondered whether the major investment in time and money into immuno-contraceptives would lead to research taking on a momentum of its own that would be difficult to stop (HRP 1993:30). Rather than critics having to justify why research should be abandoned, the onus should be on the research and funding community to explain: Who at this particular time needs antifertility 'vaccines'? By what criteria do the researchers decide whether or not to continue research on immuno-contraceptives? What mechanisms have they set up to stop, as a mimimum, the development of those methods that they themselves have classified as inappropriate?

Claim 3: a few overspeculative feminists

> Very recently, some feminist organizations have protested against research on contraceptive vaccines ... The objections to contraceptive vaccines

raised by pressure groups are, in part, due to the circulation of insufficient (or wrong) information. (Pran Talwar, in a guest editorial of *Current Opinion in Immunology* 1994b:701)

We feel the development of any new family planning method should be accompanied by honest and objective debate among all interested parties. It is a pity therefore that the authors of the document entitled 'Call for a Stop of Research on Antifertility Vaccines' ... have used what is, in our opinion, alarmist speculation, technical and scientific inaccuracies, and distortions of facts to support their opposition to the development of this new technology. (Giuseppe Benagiano, Director of HRP, 1994)

Several of the responses from the research and funding institutes to the Call for a Stop attempt to discredit critics of antifertility 'vaccines' in various ways: by casting doubt on the 'technical and scientific accuracy' of what critics say; by implying critics are a marginal minority; by questioning their legitimacy; and by setting the terms on which debate can take place.

Talwar cites two examples of the critics' 'illfounded' apprehension; the first is the concern 'that the use of these vaccines will render women permanently sterile' (Talwar 1994b:701). He asserts that none of the current products was irreversible in either animal or human trials.

His second example of an 'apprehension [which] is not valid' is a concern that '[antifertility] vaccines may be administered to women (e.g. by a dictatorial regime) under false pretences, thereby rendering the recipients unsuspectedly infertile'. He contends that 'thus far, all the vaccines in trials are fully reversible, and the protection ... is only of 3–6 months duration. Therefore, these vaccines are not suitable for such nefarious purposes' (Talwar 1994b:701).

In fact, the Call for a Stop states that immuno-contraceptives 'may last from one year to life-long', depending on the specific type. As outlined in Chapter 6, the most widely quoted anticipated duration of immuno-contraceptives – one to two years – applies primarily to anti-hCG products, although that duration has not yet been achieved. Other immuno-contraceptives, however, may have a different duration of action.

Talwar also fails to address the unresolved uncertainties over whether immuno-contraceptive users may face permanent infertility after long-term use or if they develop an overreaction. On the other hand, a duration of a few months would probably not be considered acceptable because of the likelihood of adverse effects from frequent interference with the immune system.

As there has been no pledge from researchers that they would not create an immunological contraceptive that would last longer than one to two years, it is thus misleading to limit public debate to current durations achieved.

HRP, meanwhile, has accused WGNRR and myself of spreading

'misleading statements and incomplete information'. The section in the Call for a Stop concerned with the ethics of clinical trial practices was described as questioning 'the integrity and objectives of WHO staff, consultants and investigators [which] could be considered libellous' (Griffin 1994:1, 5).[10]

Griffin also described the report launched at the 1993 Bielefeld gathering, *Vaccination against Pregnancy* (Richter 1993), as 'seriously flawed' (Griffin 1994:2) even though, prior to publication, it had been reviewed by immunologists, pharmacologists, physicians, pharmacists, a biologist, lecturers in women's studies departments, and several people in the women's and health movements. If most of these reviewers describe themselves as feminists – or at least as striving for a more just society – this does not mean that they are less professional than the contraceptive researchers and funders. It is not a *lack* of understanding that underlies the current opposition to the continued development of immunological contraceptives; it is a *different* understanding. Accusations that the critics are deliberately spreading disinformation do not further a fair and 'honest' debate.[11]

There is, however, nothing new in the attempt to discredit critics as being unscientific or anti-science. The words of Harvard Professor Emerita of Biology Ruth Hubbard are all too pertinent in the case of antifertility 'vaccines':

> Scientists tend to label as anti-science anyone who criticizes the beliefs, methods, or results accepted by the scientific elite. By drawing sharp lines, they can discredit even highly accredited scientists. Needless to say, this discreditation does not work every time, but tends to create confusion, which makes it difficult, if not impossible, for nonscientists to know whom to believe. Once confused, the public can be readily persuaded that it had best leave all the important decisions to the experts. (Hubbard 1990:209)

Other attempts at discrediting the critics, such as marginalizing them and questioning their legitimacy, make it difficult to raise public debate. David Griffin, for instance, told a Reuters journalist, '... there is an awful lot of overspeculation going on, partly from people who are opposed to the concept on philosophical grounds' (Griffin in Levene 1993:1), whilst Talwar ascribes the opposition to a few 'feminist organizations' (Talwar 1994b:701).

At the 1993 Bielefeld conference Griffin questioned how representative the opinion of the conference participants was of women in general. In response, Ellen t'Hoen, of the European office of Health Action International, advised him to go back to his Geneva office, look into the faces of his colleagues and ask them 'Let's be honest, friends, how representative are *we* to take decisions for the world's female population?' (in *Pharma-Brief* 1993:5).

Any interest negotiation is indeed a thorny issue, and representing 'women's perspectives' on contraceptives demands an enormous amount of thought, deliberation and integrity. However, researchers have a tendency to challenge and question the legitimacy of activists only when the latter's opinions are in conflict with those of the research institutions.

Legitimate activists would thus seem to be those who do not criticize the research and thus do not come into conflict. HRP, the Population Council, CONRAD and the World Bank, in their responses to the Call for a Stop, all allude to constructive 'dialogues' on immuno-contraceptives between scientists and women's health advocates.

This would seem to imply that anyone who continues to criticize the research is not being 'constructive' as defined by these institutions. It could also suggest that such 'dialogues' are sufficient and that public debates about immuno-contraceptives are thus not required.

In addition, some 'dialogues' that have taken place between scientists and women's health advocates have been misrepresented by research institutes. Talwar, for instance, despite the vociferous protests in India against the 'vaccine', made the following assertion in a medical journal:

> Dialogue with representatives of women's organizations has revealed that they are in favour of contraceptives and want more options to choose from, but they feel that more information should be forthcoming. Perhaps in the past, communication has been lacking. Women recognize the merits of the contraceptive vaccines, especially if they are further proven to be safe and effective in expanded Phase III trials. (Talwar 1994b:701)

Having attended two major HRP meetings on antifertility 'vaccines' as a consumer/women's health advocate, I for one did not gain the impression that the other women's health advocates present recognized these contraceptives' merits or were very supportive of the research. Both 'consumer representatives' invited to the 1989 HRP symposium expressed their doubts in the meeting about the appropriateness of immunological contraceptives (Hardon 1989; Richter 1992), whilst at the 1992 meeting only one of the women's health advocates present, immunologist Faye Schrater, expressed support – and that was qualified – for continued development of HRP's anti-hCG immuno-contraceptive only (see e.g. Schrater 1994b:110). The other eight women's advocates at the 1992 meeting – an immunologist, an epidemiologist, a gynaecologist, a sociologist, and consultants on women-centred family planning – either asked for more information so as to discuss with the various groups they represented or opposed further development. All expressed concerns about unresolved aspects of the research (HRP 1993:31).

At another HRP co-sponsored meeting in March 1993 in Mexico, the International Symposium on Contraceptive Research and Development for the Year 2000 and Beyond, the vice-president of the International Women's Health Coalition, Adrienne Germain, recommended that 'no further work be done on the contraceptive vaccine so that the resources – human as well as financial – can be moved to priorities higher in women's agendas' (Marcelo and Germain 1994:338).[12]

At the same gathering, the head of the Population Program of the MacArthur Foundation, Carmen Barroso, called on scientists and policy makers not 'to be discouraged by the difficulties that a dialogue between women and scientists may pose', pointing to an 'honest assessment of the obstacles' as being an indispensable precondition for fruitful debates. 'If we are committed to fostering dialogue between women and scientists, we should be ready to face its wide implications in the difficult process of shifting resource allocations' (Barroso 1994:518).

'Dialogues' – which I prefer to call 'interest negotiations' – can indeed have a role in democratic decision making on technology development. Yet, as Betsy Hartmann says, 'Dialogues have sometimes muted dissent and served to legitimize the contraceptive research agenda without altering it in any significant way. Words are cheap. Action is not' (Hartmann 1995:284).

Those institutions that refer to 'constructive dialogues' should take note of a passage in HRP's report of the 1992 meeting which emphasized:

> The women's health advocates expressed concern about their role in the meeting. They were afraid that their names and presence would be used to legitimize both the content and the process of research in an area about which many were doubtful. (HRP 1993:28)

Claim 4: increasing women's choices Assurances that there is no need to worry about the biomedical shortcomings of immuno-contraceptives or their potential widespread abuse coupled with references to more constructive 'dialogues' with women's health advocates are all part of a 'women-friendly image' that population institutions have striven to create over the past decade or so.

Thus one of the most pervasive characteristics of the institutions' responses to the Call for a Stop is the assurance that these institutions are fervent supporters of 'free and informed choice' for women in the matter of reproduction. The president of the Population Council, Margaret Catley-Carson, made this assertion:

> The Population Council is dedicated to advancing the reproductive health of women ... The mandate of the Population Council's contraceptive research and development program is to increase the number of safe and effective contraceptives available. We develop contraceptives to enable women and men to regulate their own fertility in accordance with their own goals. (Catley-Carson 1994:1)

The USAID-financed Program for Contraceptive Research and Development (CONRAD) has declared its 'commitment to expanding the free and informed choice of women and men to regulate their fertility as safely and in as user friendly fashion as possible' (Gabelnick 1994:1). HRP had said it is striving for the provision 'of a wider choice to users by developing new and improved methods of fertility regulation that are safe, effective and acceptable' (Benagiano 1994:1), while a major funding organization, USAID, asserts that it 'has had a longstanding commitment to expanding contraceptive choice through the provision of a broad range of options to meet diverse user needs' (Gillespie 1994:1).

HRP director Giuseppe Benagiano, moreover, wrote in his letter:

> I agree completely with the aim of Women's Global Network for Reproductive Rights … namely the 'right of women to decide whether, when and how to have children'. It is, however, my contention that this aim also includes the right of women to choose what method of family planning to use, including, if they wish, an antifertility vaccine.

Faced with statements such as these, one might easily construe that any women's health and rights activist who raises concerns about the contraceptive 'vaccine' must be *against* women's 'choice'. However, at the 1992 HRP meeting, women's health advocates emphasized that 'increasing the number of methods available does not automatically lead to expanded choice'. They cited concrete examples of how different interests shaped the methods that were in practice available to women in different settings (HRP 1993:23).

Choices are shaped by what is available and the power to choose. As community health researchers Rani and Abjay Bang report from India:

> In reality, the choice of contraceptive methods is not made by women. The decision is actually often made by the government health programme officials and workers. Any new method is pushed aggressively without bothering to evaluate and learn from the failures of the previously propagated methods. (Bang and Bang 1992)

Indeed, the increasing of reproductive 'choice' cannot be understood as the right of research institutions to develop whatever they consider to be attractive and feasible, nor should it be seen as a 'choice' between immuno-contraceptives of different durations to the exclusion of a prior question: should certain technologies be developed at all?

In the words of Annette Will from the BUKO Pharma-Kampagne:

> The introduction of one more family planning method does not give people more or less freedom to choose but more or less things to choose from. Nobody deprives women of the freedom to choose, when one objects to one

or more (bad) contraceptives to chose from. Furthermore, more choice has no meaning in itself; what is important is the question: more choice of what? ... Reproductive self-determination ... is not a question of developing another control device, but it is a complex social and political issue that affects men and women differently. (Will 1995:2)

Because the shaping of women's 'choices' begins with decisions over what is actually developed or not developed, the Call for a Stop is explicit about the 'demographic-driven, science-led'[13] framework in which the concept of antifertility 'vaccines' has been developed and which defines women's choices:

The major funders of contraceptive research want to increase the effectiveness of population control programmes. Most of the scientists involved have been taught to see the body as a machine.[14] The major trend in contraceptive development has been to create technologies which are long-acting, have a low 'user failure' rate, which lend themselves for mass fertility control – and which interfere with delicate and complex processes in the human body. (Appendix 3)

Although some researchers have stated that contraceptive 'vaccines' have to be developed because of 'overpopulation' (see Chapter 5), not a single response to the Call for a Stop referred to the usefulness of antifertility 'vaccines' in stemming population growth, or to their major theoretical advantage of having a low risk of 'user failure' rate.

Most responses avoided the term 'population' altogether, while a few tackled the debate head on. Talwar professes that 'women's concerns for empowerment, of having the right to decide on their reproductive status and not be treated as machines for attaining demographic targets, are legitimate, and have to be respected' (Talwar 1994b:701).

'We believe family planning should be a strictly voluntary service, free of demographic and provider targets, to assist individuals in reaching personal fertility goals healthfully,' affirms Population Council president Margaret Catley-Carson, referring interested people to the Population Council's new pamphlet *Population Growth and Our Caring Capacity* for a 'fuller explanation of the institution's approach to population and related issues' (Catley-Carson 1994:3). This pamphlet states that:

Population policies designed to accelerate voluntary fertility decline have increasingly been equated in the public mind and in the budgets of policy makers as fertility reduction implemented through family planning programs. This approach is inadequate and, in some cases, counterproductive. Some family planning programs have been misshapen by implicit or explicit demographic imperatives while the vast potential of progressive, social, economic and health measures to affect population growth [has] been underplayed. (Quoted in Population Council 1994:5).

It seems from this as if the population institutions have finally taken on board the concerns of women's activists. But as indicated in Chapter 5, the change has largely been one of rhetoric or, at most, one from 'hard' to 'soft' population control.

Catley-Carson's assertion is, in particular, an ahistorical view of the Population Council's role in contraceptive development.[15] Any assurances that antifertility 'vaccines' are being developed in a framework centred solely on increasing women's voluntary informed choices seem to have forgotten the prevalent thinking when research began.

For instance in 1969 in a widely publicized article in *Science*, the Population Council's then president, Bernard Berelson, reviewed proposals to limit population growth more effectively. Suggestions for 'involuntary fertility control' included 'temporary sterilization of all girls by means of time-capsule contraceptives, and of girls and women after each delivery, with reversibility allowed only after government approval' and 'mass use of a "fertility control agent" by the government to regulate births at an acceptable level ... [through] a substance now unknown but believed to be available for field testing after 5 to 15 years of research work. It would be included in water supply in the urban areas' (Berelson 1969:533).

Berelson concluded that 'involuntary fertility control' was ethically doubtful – and not the most effective means of population control. He warned that as coercion creates resistance, more subtle means should be employed to get people to control their fertility.

Critics have pointed out that the characteristics of Norplant® fit exactly the specifications of the first proposal for 'involuntary fertility control' – temporary sterilization by means of time capsules.[16] How close could oral antifertility 'vaccines' come to the second suggestion – mass use of a fertility control agent?

The Population Council's silence about its role in developing contraceptives that are prone to abuse and its active role in spreading population control ideology and programmes worldwide, while at the same time blaming others for pursuing demographic goals, lays the responsibility for contraceptive abuse at the feet of others while absolving itself.[17]

In her rejoinder to Margaret Catley-Carson, Annette Will comments that a change in the overall ideology of the Population Council from a demographic to a more 'women-friendly' approach would be easier to believe 'if you would address your own history' (Will 1994:3).

Were organizations such as the Population Council to address their own history, they would make it easier to discuss the impact that population and other ideologies have had on contraceptive design, clinical trials and service delivery. To insist that these matters need to be brought into a public debate is not tantamount to an allegation that researchers plan abuse of antifertility 'vaccines' or do not care about the health of future users (although it does not exclude these issues). It is primarily a call for the research and funding community to be reflective and to clarify the frameworks of thought that may have influenced their judgement and practices in the past and in the present – and to demand that they take responsibility for the predictable consequences of their work.

Many critics feel that this debate is of prime importance. For the campaigners, the framework of present contraceptive research is clear:

> Population control ideology should not guide the development of contraceptives. The aim must be to enable people – particularly women – to exert greater control over their fertility without sacrificing their integrity, health and well being. Contraceptive development must be oriented at the realities of women's lives. Above all it must consider local health care conditions and the position of women in society. (Appendix 3)

Glimmers of hope[18]

> By the time a technology is sufficiently well developed and diffused for its unwanted social consequences to become apparent, it is no longer easily controlled. Control may still be possible, but it has become very difficult, expensive and slow. (Collingridge 1980)

When people first hear about the campaign calling for a stop to research on immunological contraceptives, they tend to ask 'Yes, but can we actually stop research?' Many critics of the method believe that it is important at least to try and influence the agenda of the major research institutions and their funders, and indeed, by the middle of 1995, some changes could be detected.

Information about antifertility 'vaccines' has been circulated and the topic has been discussed at many conferences worldwide, so that various health professionals and others are beginning to speak out against further development. In the run-up to the UN International Conference on Population and Development, held in September 1994 in Cairo, antifertility 'vaccines', in particular their technology-inherent potential for abuse, emerged as a recurrent theme in the NGO preparatory meetings.

For example, 215 women from 79 countries who attended the Reproductive Health and Justice International Women's Health Conference, co-organized by the International Women's Health Coalition and the Brazilian Citizenship Studies Information and Action Organization (CEPIA), recommended that 'resources should be redirected from provider-controlled and potentially high-risk methods, like the vaccine, to barrier methods … female-controlled methods that provide both contraception and protection from sexually transmitted diseases including HIV, as well as male methods' (IWHC and CEPIA 1994:6).[19] Those endorsing this recommendation included representatives of organizations and networks active in the field of health, human rights, development, environment and demography.

A network of Southern women activists and researchers, DAWN – Development Alternatives with Women for a New Era[20] – gave a similar recommendation in its policy report on Population and Reproductive Rights launched in September 1994 at the UN Population Conference (Corrêa 1994:109).

One of the first health professionals to declare his reservations about this line of research was Graham Dukes, editor of the *International Journal on Risk and Safety in Medicine*, who believes that the balance between the risks and benefits of immuno-contraceptives in the 1990s is as unfavourable as it was years ago and that their development is 'asking for trouble' because their adverse effects were difficult to predict in tests but could be 'very serious' (in Wieringa 1994:2; Dukes 1995).

An official with the Swedish drug regulatory agency, Arne Victor, stated in June 1994 at a public meeting in Sweden[21] that the technology-inherent features of antifertility 'vaccines' were such that, once available, no arrangement could guarantee the social control of them necessary to prevent their abuse.

Victor said that since he had voiced his opposition to the development of antifertility 'vaccines' way back in 1989 at the HRP Symposium on Fertility Regulating Vaccines, he had tried to raise the issue in Sweden, but that neither fellow scientists nor the media had shown much interest.

In Norway, however, the *Arbeidersbladet* newspaper has quoted one critic, Staffan Bergström, gynaecologist and professor at the Oslo Centre for Development and Environment:

> We should not be hostile to technology, but I am afraid that the antipregnancy vaccine will be part of the 'coercive methods' that are used against women in Third World countries to make them have fewer children ... Contraceptive devices that women themselves are not able to take out or remove deprive them of the possibility of self-determination. For women in Third World countries the possibility to decide for themselves is important because they often have little access to health services and follow-up services. (in Egjar 1994)[22]

Around thirty women physicians from all over the world who participated in a workshop on immuno-contraceptives during the 23rd Congress of the Medical Women's International Association in May 1995 unanimously passed a recommendation that the association should use its influence with the WHO and other organizations to stop the development of antifertility 'vaccines' (Fuente 1995).[23]

Meanwhile in India, at a plenary lecture entitled 'Accountability in

Medical Research' which took place at the Science Congress in Jaipur in 1994, A. S. Paintal, a former director-general of the Indian Council of Medical Research, launched a scathing attack on the 'political patronage of unscrupulous scientists' which he described as 'one of the most unfortunate things that afflicts Indian science'. He said that 'scientists seem to get everything done through lobbying and maintaining public relations'. Paintal called attention to the potential long-term risks of the cross-reacting hCG antigen used in the NII's Phase II (efficacy) trial, and advocated that 'the authorities concerned in India must terminate or at least suspend these trials'. He concluded:

> A bad precedent has been set by exonerating guilty scientists in the past. The Indian authorities now seem powerless in taking punitive action against individuals guilty of gross scientific misconduct. (in Hari 1994:96, 98; *Times of India* 1994)

Funders under scrutiny   Halfway across the world, representatives of one of the NII's major financial sponsors for the past two decades, the Canadian International Development Research Centre (IDRC), voiced a markedly different opinion to Paintal when they met thirty women activists in June 1995. The women were attending the Second International Action Workshop on Antifertility 'Vaccines' (the first being the Bielefeld gathering in 1993) and had requested a meeting with the IDRC.[24]

Two days before the meeting, the IDRC released a position paper on antifertility 'vaccine' research which it circulated among the mass media (IDRC 1995a and 1995b)[25] and which was, according to an internal memo, intended 'to contain negative publicity'. (The activists were only given copies at the start of their meeting with IDRC officials.) Judging by this paper, it would seem that the driving force behind the IDRC's decision to fund the NII's research had been the desire to 'widen the options for contraceptive choice for women' and to enable a woman to 'better protect her own health and the well-being of her children' (IDRC 1995a:1, 2 and 1995b:1). 'Extensive animal trials ... indicated no adverse effects,' the paper stresses, and 'to date, there have been only minor side effects'. This is attributed to the fact that 'this method of birth control causes less interference [with] natural functions than hormonal based contraceptives' (IDRC 1995b:1, 2). The position paper maintains that there is no need to worry about 'dangers to the immune system':

> The human system works constantly to protect us from a variety of diseases and infection. Advance stimulation of the immune system through vaccination is a common practice that has proven of great benefit to humanity. As with any vaccine, there are potential risks. These are being watched for and they

have not manifested themselves either in animals or human trials. The risks
with this contraceptive vaccine are minimal because the levels of reactive
substances involved are very low (50 parts per billion) and not associated with
body tissues such as organs. (IDRC 1995b:2)

In one section on duration, the position paper states that the immuno-
contraceptive 'lasts about one year', while in another section two pages
later covering the 'potential for abuse' it states that 'since vaccines have
a duration of effectiveness of *less than one year* and are completely
reversible, they are less likely to be open to abuse than sterilization and
other long-acting contraceptives' (IDRC 1995b:2, 4, my emphasis).

The paper maintains that reversibility on demand is not a problem
because 'The effect of the vaccine could also be switched off at any
time, either by administration of hCG to absorb the antibodies or by
taking progesterone' (IDRC 1995b:2).

It states that the 'IDRC is committed to conducting research only
under strictly rigorous scientific and ethical rules'; as proof of this, a
detailed account of the ethics review procedure is cited (IDRC 1995a:1).
Doubts about the enrolment procedure in the Phase II (efficacy) trials
are dismissed:

A group opposed to the vaccine have produced a video [*Antibodies against
Pregnancy* by Ulrike Schaz] apparently showing a sequence where women
appeared to be recruited without appropriate informed consent. We believe
that there is a misunderstanding about what was actually being portrayed.
This sequence is likely a pre-screening at a family planning clinic where
women were being identified as possible potential candidates who later had
the choice to enter into the trial. Those interested then had to go through
[an] elaborate consent procedure ... All women who eventually participated
in the trial gave signed informed consent. (IDRC 1995b:3)

Although the IDRC has a reputation as a progressive funding agency,
this position paper – and two constantly clicking digital 'clocks' in the
IDRC's lobby which compare growing population with declining arable
land – cast some doubts in activists' minds about the institution's
women-centred approach. The position paper might have appeased the
media, but if it was intended to allay concerns raised by critics about
immuno-contraceptives, it did not; instead, it did the reverse.

When the activists met for two hours with five IDRC officials, neither
the IDRC's director, Keith Benzanson, nor the IDRC official for
biomedical research, Don de Savigny, could explain why the IDRC had
supported the NII's research with more than $CAN4.5 million and is
sharing the patent rights on the product, despite the controversies over
the ethics of Talwar's research which were first raised in the mid-1970s
(see p. 105) and despite the US Food and Drug Administration's advice
against conducting the Phase II (efficacy) trial using the NII's doubly

cross-reacting anti-hCG contraceptive (see pp. 25–6, 97) – other than to mention the historical relationship between the IDRC and Talwar over the past twenty years.[26]

Contrary to their position paper, the IDRC officials were able to provide scientific proof neither of how the anti-hCG contraceptive could be safely 'switched off' nor that the duration of action of the current prototype was one year. Nor did Don de Savigny maintain his initial view that the prototype's efficacy rate of 80 per cent could be considered a 'success'.[27]

The activists requested public clarification of those passages in the position paper that could easily be misunderstood as implying that the quality and ethical conduct of the research, and the efficacy and safety of immuno-contraceptives had been proved.[28]

They pointed out that the report of the NII's Phase II (efficacy) trial, which according to the IDRC 'continued to confirm the absence of adverse effects' (IDRC 1995b:1), did not contain any data of nor mention adverse effects. Had the adverse effects recorded in nearly one-third of the Phase I (safety) trial participants (Talwar et al. 1990a) simply disappeared? (See Chapter 3.) Even if the investigators had not reported any serious adverse effects from Phase I or Phase II, this should not be interpreted as a reassurance of the safety of an experimental product. Even after years of being available, some medicines have been withdrawn from the market because of their lack of safety.

The comparison between immuno-contraceptives and anti-disease vaccines seriously clouds the issues. As the antibody concentration of 50 parts per billion is deemed sufficient to have a contraceptive effect, there is no reason why it should not also be able to cause immune-mediated adverse effects.

The IDRC's contention that the women filmed in *Antibodies against Pregnancy* who were being enrolled on clinical trials had given their 'informed consent' was based on assurances from NII researchers that they had followed the rules governing clinical trials and a number of women had signed consent forms.[29] One Indian health activist thus asked the IDRC officials whether they believed that those women who were filmed being given assurances of the safety and efficacy of antifertility 'vaccines' were subsequently told that such statements were all a mistake and that actually the aim of the trials was to discover what the risks and contraceptive benefits of this totally new contraceptive product were.

It is relevant to note that the Council for International Organizations of Medical Sciences makes the following specification in its International Ethical Guidelines for Biomedical Research Involving Human Subjects:[30]

Obtaining informed consent is a process that is begun when *initial* contact is made with a prospective subject and continues throughout the course of the study. By informing the subjects ... by assuring that each procedure is understood by each subject, the research team not only elicits the informed consent of the subject but also manifests deep respect for the dignity of the subject. (CIOMS 1993:14, my emphasis)

Film-maker Ulrike Schaz was one of those present at the meeting with the IDRC and explained that, when filming, she herself had been taken aback when the explanations given to the women she filmed about the 'vaccine' were translated for her. Schaz explained that although she was critical of the development of antifertility 'vaccines', she had not expected this kind of enrolment procedure in a major New Delhi hospital. To check whether these were exceptional cases or not, Schaz asked a trial investigator how she was able to tell the women that there were no side effects. The investigator told her that this was the information she herself had been given, and that women would tend to follow the advice of the doctors.[31]

As a result of this meeting, the IDRC officials promised to circulate to the mass media a statement corrective of its position paper. They agreed to provide a copy of the ethics committee reports, list the criteria by which IDRC decided to fund contraceptive research in the 1970s and the 1990s, and to involve a representative from one of the Indian groups present in an independent review of the ethics of the NII trials, the informed consent procedure, and the monitoring of the long-term health effects on the trial participants and their children.

However, the statement the IDRC's Director of Public Information Program sent a few days later to one of the Canadian co-organizers of the meeting did not so much rectify inaccuracies in the IDRC's position statement as provide 'additional information' (Spierkel 1995).

Other funders have been more open to discussions about the appropriateness of developing immuno-contraceptives. For instance, after a parliamentary inquiry into the Dutch government's contributions to HRP's immuno-contraceptive research, the Netherlands Minister of Development Cooperation, Jan Pronck, convened two small consultative meetings; he invited a few health professionals, development consultants and government officials to discuss the issue with, at the first meeting, women's rights and health advocates and, at the second, two representatives of HRP. In June 1995 at the second meeting, the minister stated that 'as a policy maker' he could not ask for research to be stopped permanently (although he did not completely discount a moratorium), but preferred instead to use his influence to ensure the best possible research practice in publicly funded institutions.

Changes in research criteria? There has also been some indication of possible changes among research institutions. In June 1994, for example,

HRP convened a Discussion on the Ethical Aspects of Research, Development and Introduction of Fertility Regulating Methods. A considerable part of this meeting, which brought together several specialists in the area of women's health advocacy and human rights and members of HRP's Scientific and Ethical Review Group (SERG), was devoted to attempts to define contraceptive abuse.

The Australian investigator of HRP's Phase I (safety) trial, Warren Jones, and I were invited to address 'specific considerations for antifertility vaccines'. We were each given fifteen minutes to elaborate on whether antifertility 'vaccines' would have any advantages over other contraceptives – and, if so, whether their potential for abuse would offset these 'perceived' advantages. Needless to say, there was little common ground between the two presentations.

Another main topic discussed at this meeting was 'the rationale for selecting methods for research and development' (HRP and SERG 1994:2), that is, the way in which contraceptive research priorities are decided upon. Some participants were surprised to learn that HRP's list of criteria for selection – including *demand* defined as an 'explicit request from member states (World Health Assembly, Ministries of Health, donors, other intergovernmental organizations)', *need and rationale* 'as perceived by the scientific community', and *applicability* defined as 'the extent to which the results can be expected to have an impact on family planning programmes and practice'[32] – dated back to 1979 when issues such as AIDS and the differential gender impacts of policies were not considered.

The meeting recommended that HRP should revise its framework for establishing research funding priorities to include 'an assessment of abuse potential and consideration of health service infrastructure, gender-based power imbalances, and pattern of disease, especially HIV and other STDs' in its 'risk/benefit analysis' (HRP and SERG 1994:7).[33]

If this recommendation were to be taken seriously by HRP and it did change its official definition and assessment of contraceptive 'risks' and 'benefits', such a shift would have important implications for funding and research decisions in general, not just within HRP.

Can we stop research?

The wide variety of signatories of the petition Call for a Stop of Research on Antifertility 'Vaccines' (Immunological Contraceptives) – ranging from a multitude of women's organizations to health, human rights and consumer action groups, from alternative development policy groups to aid agencies, student associations and workers' unions – shows that women and men worldwide are worried at the direction contraceptive research has taken.

The campaign against antifertility 'vaccine' research has not sparked this concern. Much of it is a consequence of the long-standing efforts of critics both without and within the contraceptive research community who believe that there is a need to strive for more democratic decision making in developing new technologies. An important difference between previous campaigns focused on birth control methods that are not felt to be women-centred (such as Norplant® and Depo Provera®) and the campaign on immuno-contraceptives is that the 'vaccines' are still under development and some years away from being approved by drug regulatory authorities.

Annette Will of the BUKO Pharma-Kampagne urges that this lead time must not be wasted:

> We must not postpone reflecting and evaluating a new technology in all its potential consequences to a later point of time. History teaches us that there is no such thing as a neutral technology or a neutral science. Scientific research is always carried out within a social, political, religious, cultural and economic framework … Therefore for us the leading questions are: *Why* are immunological contraceptives being developed? *For whom* are they meant? *By whom* are they researched? *Who* has *which* interest in the development of immunological contraceptives? What will they do to women and men? How are they going to influence people's health, dignity and integrity? (Will 1995:2, original emphases)

No one can tell at this stage if it is possible to halt research into immunological birth control methods and to redirect contraceptive development towards being truly people-centred. But determination, networking and vision are important ingredients in the worldwide struggle for birth control as a right – not a duty.

Notes

1. My brief descriptions of some of the actions and public statements of concerned citizens and groups prior to and during the campaign to raise public awareness on antifertility 'vaccines' are inevitably incomplete. I have included the main ones I am aware of.

2. This is the meeting that HRP held over 17–18 August 1992 in Geneva (HRP 1993).

3. The BUKO Pharma-Kampagne and the HAI-Europe coordination office also invested considerable staff time.

4. There were three abstentions: from David Griffin, from a medical student, and from a consultant of GTZ, the German agency for technical cooperation.

5. The twelve countries represented were Argentina, Chile, Dominican Republic, Egypt, Finland, Germany, India, the Netherlands, South Africa, Switzerland, Zambia and Zimbabwe.

6. The Action Workshop was organized by the BUKO Pharma-Kampagne, the Amsterdam-based Women's Health Action Foundation (formerly the

WEMOS Project on Women and Pharmaceuticals) and the coordinating office of the Women's Global Network for Reproductive Rights (WGNRR).

7. For more details about the dynamics of contraceptive research, in particular the reluctance of scientists to engage in research that provides low scientific status, such as the improvement of barrier methods, see Clarke 1995; Wajcman 1991; and Wajcman 1994. The arrival of AIDS and long-standing pressure from women's health activists have changed this agenda somewhat. For a feminist critique of science see Harding 1991.

8. The muscle and joint pains were attributed by the researchers to the adjuvant. They are the same type of adverse effects that led to the suspension of the Phase 2 (efficacy) trials in 1994 (HRP 1994b:8).

9. For a first answer to the article of Griffin and colleagues, see Richter 1994b. (The fully referenced article can be requested from the Women's Global Network for Reproductive Rights.)

10. David Griffin wrote, 'To suggest, as you have done in your flyer entitled "Call for a Stop" that WHO has conducted unethical clinical trials in this area is clearly a statement at variance with the fact. Furthermore, it questions the integrity and objectives of WHO staff, consultants and investigators and could be considered libellous.'

11. The topic of antifertility 'vaccines' is still relatively new and continuously evolving, while the research material is often not readily accessible to those outside the research community. I would not, therefore, exclude the possibility that there are unintentional errors in this publication. Any comments and suggestions for improvement are more than welcome.

12. In their presentation on 'Women's Perspectives on Fertility Regulating Methods and Services', Marcelo and Germain stated: 'With regard to the methods currently under investigation, one that worries many women's health advocates is the contraceptive vaccine. This is not the place to present a detailed argument. Suffice it to say that many women have grave reservation about the concept, the scope of unanswered questions, even after two decades of the research, its potential for abuse by the state, and its possible consequences for other social programmes (such as child immunization). While some women support further research to answer their questions, many others, including the authors, recommend that no further work be done on the contraceptive vaccine so that the resources – human as well as financial – can be moved to priorities higher on women's agendas.'

13. David Griffin's description at the June 1993 Bielefeld meeting.

14. The original sentence was: 'Researchers see the body as a machine to be mastered.' The sentence was later amended because some campaigners felt that it gave the impression that researchers intentionally set out to master. They also felt that, rather than pointing to a reductionist Western science model, the phrase could be misconstrued as being anti-science. For this book version I have also deleted the claim that the effects of the Indian formula 'lasted from less than three months to over two years', having seen NII's new phase II report (see note 8, Chapter 3).

15. Pramilla Senanayake and Janet Turner of the International Planned Parenthood Federation summarized the history at the HRP co-sponsored 1993 Symposium on Contraceptive Research and Development for the Year 2000 and

Beyond as follows: 'When John D. Rockefeller III founded the Population Council in 1952, none of the modern methods of family planning existed. The very notion of family planning was extremely sensitive and it had little public support anywhere in the world, including the USA. Rockefeller was convinced that too rapid population growth constituted a threat to people and nations, and he formed an independent, non-profit organization to study the interconnections between demography, reproductive physiology, and family planning in the developing world. The Council established its biomedical research laboratory in 1956 at the Rockefeller Institute for Medical Research (now the Rockefeller University) in New York City. The Council's work focused on the development and testing of contraceptives that were seen to be particularly suitable for use in developing countries' (Senanayake and Turner 1994:253). For a critical account of the histories of the Population Council and other agencies, see e.g. Hartmann 1995:113–24.

16. In the words of medical anthropologist Soheir Morsy, the hormonal implant 'harbours an inherent potential for the imposition of biomedically-mediated social control over women's reproductive capacities'. According to Morsy, ensuring that specific types of women use Norplant® does not necessitate outright coercion; it can be achieved through much more acceptable ways of 'social engineering'. Rather than being assisted to make an informed choice about their method of fertility regulation, women can be 'persuaded' that Norplant® is best for them (Morsy 1993:91).

17. 'The Population Council has played a crucial role in implementing population control programmes in the Third World countries, which misused women as targets for demographic purposes,' according to health activists Sumati Nair and FINRRAGE International Coordinating Group member Ulrike Schaz (Nair and Schaz 1995:1, 2). 'The Council had an important impact in legitimizing population control politics by giving scientific status to demography and the ideology of "overpopulation". This helped to establish an extremely ahistorical, reductionist and racist view to problems and people in Third World countries. The Population Council is silent about its political roots and its involvement in population control. In your letter [to WGNRR in response to the Call for a Stop] you blame Third World governments for the manner with which the contraceptive is promoted within a country. The final accountability for Norplant®'s spread and incorporation for use in more and more Third and now First World countries, does lie with the Population Council.'

18. Thanks to Betsy Hartmann for inspiration for this title (Hartmann: 1995:279).

19. In the past, discussions about the appropriateness of various contraceptives have tended to focus on the extent of their 'provider dependency'. This term, however, confuses several issues. Description of a contraceptive as 'provider-dependent' only points to the fact that a provider has to administer the contraceptive (such as hormonal and immunological injectables, implants and IUDs) and remove it (implants and IUDs). The term fails to address fully the issues either of potential abuse or of the degree to which the safety of individual methods (particularly those that profoundly affect our bodies but cannot be discontinued at will) depend on a good medical back-up system. For example, oral immuno-contraceptives acting against sperm, which could potentially have

a lifelong effect, would not be 'provider dependent' – a woman could take them by herself. Yet they could be one of the most abused methods and one that, if there were adverse effects, could not just be 'switched off' by the user. This is why I would propose abolishing the term 'provider dependency' and clearly distinguishing between the aggregate of technology-inherent features which may foster abuse and those which determine the need for a good medical back-up system.

20. DAWN is a network of some 4,500 Southern women activists and researchers concerned with the impact of 'development models' on gender systems and on women's socioeconomic, political and cultural situation.

21. The meeting was co-organized by Health Action International, Sweden, and Kilen, a Swedish support and public awareness-raising centre on tranquillizer abuse.

22. Thanks to Trine Lynggard, CEWI, Norway, for translation.

23. The recommendation was not adopted by the General Assembly of the association because this was the first time that most participants had heard about the research into antifertility 'vaccines' and they wanted more information.

24. The Action Workshop was co-organized by Women's Health Interaction, Ottawa; the Committee on New Reproductive Technologies of the National Action Committee, Toronto; and the Women's Global Network for Reproductive Rights' Coordination Office, Amsterdam. The participants were from Australia, Brazil, Canada, Dominican Republic, Germany, India, Indonesia, Kenya, Mexico, the Netherlands, Nigeria, Peru, the Philippines, the USA, and Zimbabwe. The major aims of the meeting were to decentralize the campaign further and to map out new strategies – not only to stop the research on antifertility 'vaccines' but also to redirect contraceptive research. The participants also decided to change the campaign name to 'Stop Antifertility "Vaccines": International Campaign Against Population Policies and Hazardous and Abusive Contraceptives' to emphasize its broad political focus.

25. The IDRC's position paper had three parts: 'IDRC's Position' (IDRC 1995a), 'Most Commonly Asked Questions on the Contraceptive Vaccine' (IDRC 1995b) and 'IDRC's Work in Women's Health Issues - An Overview' (IDRC 1995c)

26. An IDRC anthropologist, Jennifer Loten, described the IDRC's involvement in population issues as follows: 'When IDRC was established by an Act of Parliament in 1970, the world looked at the process of development and the challenges posed by issues such as population through a very different lens. Population control and fertility regulation were embraced as solutions to a problem defined in terms of crisis. The population "explosion" was described as a crisis of global proportion, economically, ecologically and politically. IDRC was not the only organization to locate their approach in the health sciences disciplines, focusing on fertility regulation. So close was the association between population and the human body that the first health division at IDRC existed as Population and Health Science. A full 50% of the programme's budget was devoted to fertility regulating projects' (Loten 1993:3). Noting that the IDRC's first grant to Talwar was granted in 1974, Loten goes on to describe the changes that came about as a result of the 1974 World Conference on Population: 'Population lost its demographic cast as it matured in Social Sciences, and Health Science

came to link family planning and fertility regulation more and more with community health' (Loten 1993:5). 'In 1991, the closure of [IDRC's] Population, Education and Society programme signalled the disappearance of population as an explicit programme within the Social Science division. Far from abandoning the issue, however, it is part of the division's discourse implicitly as an under-pinning to all elements of its research' (Loten 1993:6).

27. Although Vernon Stevens has stated that 'something with less than 80 per cent response is not acceptable in the West' (in Stackhouse 1994:A2), it is not yet clear whether it would also be considered unacceptable in India.

28. The same criticism of the research and funding community was made at the 1992 HRP meeting between women's health advocates and scientists when 'it was agreed that the words "safe", "effective" and "acceptable" are often used too liberally in connection with drugs and vaccines which have been tested in a limited number of animal studies and on a very small number of people. In the case of the anti-hCG vaccines, some researchers have claimed that they are "safe" and "effective" on the basis of results obtained in Phase I or II clinical trials and it was felt that this was misleading' (HRP 1993:26).

29. Maureen Law, director-general of IDRC's Health Science Division, wrote to Canadian activists on 18 May 1995, prior to the June meeting between activists and IDRC, that: 'According to the clinical trial protocol, recruitment of volun-teers to the trial includes an informed consent procedure consisting of a local language information package and informed consent document containing information on the full nature of the trial and the vaccine, risks and benefits ... etc. The informed consent document is signed by the participant, the local investigator, and a witness, and a copy remains with the participant. Participants were to be literate and urban, mostly recruited from nurses on hospital staff. After the trial was completed, there was an allegation [i.e. in the film *Antibodies against Pregnancy*] that there may have been irregularities in the information provided in the informed consent process at one recruiting point. This has been taken up with the researchers and it is our understanding that it has been agreed that a witness from a local women's NGO would be present at all future clinical trial recruitment' (Law 1995:2, 3).

30. The CIOMS guidelines are an internationally recognized supplement to the Helsinki Declaration on ethics in clinical trials. They were established to deal, for example, with ethical dilemmas caused by the internationalization of research. This includes the question of accountability of sponsoring agencies. At the 1989 HRP Symposium on Fertility Regulating Vaccines, all research and funding agencies were called upon to observe the CIOMS guidelines. (See Report 1991:280.)

31. In a taped interview with Ulrike Schaz, the investigator said: 'The women are very trusting, they don't ask too many questions. They come with faith to the doctor. So if we say "It's safe and it's good for you – rather than having two Copper-T insertions for six years, change to this for a year, then you can switch back" – they do it. They go by what we advise.'

32. 'Criteria Used by the Programme in Setting R&D Priorities' (HRP/AD.GR/1979); overheads 17–19 of the presentation by HRP associate director Paul van Look on 'The Rationale for Selecting Methods for Research and Development'.

33. For other topics discussed and recommendations made, for example, on how to prevent and document practices of disinformed consent in clinical trials, see the summary report of the meeting (HRP and SERG 1994). However, the discussions did not always distinguish between problems of voluntary informed consent in clinical trials, abuse of contraceptives, and lack of access or selective access to contraceptives.

HRP meetings on immuno-contraceptives

Evaluating the Safety and Efficacy of Placental Antigens for Fertility Regulation, 8 February 1978

Document drawn up by the HRP's Task Force on Immunological Methods for Fertility Regulation in consultation with specialists in immunology, toxicology, immunotoxicology, immunobiology, reproductive biology and representatives of three national drug regulatory authorities – to serve as a model guideline for safety and efficacy testing of anti-hCG contraceptives.

Symposium on Assessing the Safety and Efficacy of Vaccines to Regulate Fertility, 12–16 June 1989, Geneva

Primary objectives: to review the principles of antigen selection, draw up guidelines for further trial phases (with a particular emphasis on anti-hCG immuno-contraceptives), and discuss the social, legal and ethical issues that may be raised by the development and use of immunological birth control methods.

Participants: more than fifty attended, including around thirty scientists (among them representatives of the five major research institutions), seven industry representatives, four representatives of drug regulatory authorities, two lawyers, a few social scientists and two consumer advocates.

Meeting between Women's Health Advocates and Scientists to Review the Current Status of the Development of Fertility Regulating Vaccines, 17–18 August 1992, Geneva

Primary objective: to 'initiate an open exchange of information, opinions and ideas between scientists and women's health advocates about this fundamentally new approach to fertility regulation'.

Participants: ten scientists who were involved in either conducting or funding research on immuno-contraception (from India, Europe, and the USA) and ten women's health advocates (from Africa, Asia, South and North America, and Europe) with backgrounds in immunology, health service delivery, and social and clinical research, as well as wide experience in working with women in health-related areas.

Discussion on Ethical Aspects of Research, Development and Introduction of Fertility Regulating Methods, 1–2 June 1994, Geneva

Organized by the HRP in conjunction with its Scientific and Ethical Review Group (SERG).

Primary objective: 'to review ethical issues related to research, development and introduction of birth control methods, with a particular emphasis on the potential for abuse'. Four main areas were addressed: priority-setting in contraceptive research; abuse potential; the specific case of antifertility 'vaccines'; ethical practices in research, introduction and widespread use of birth control methods.

Participants: seven women's health and/or rights advocates, the principal investigator of HRP's phase I trial, Warren Jones, and nine members of HRP's ethics committee.

Chronology: research on – and resistance to – antifertility 'vaccines'

Turn of the century

Scientists discover that injection of sperm or testes extracts into animals can induce immune reactions against sperm.

1920s to mid-1930s

A first wave of human trials occurs testing 'spermatotoxins' in women.

1939

The US National Committee on Maternal Health advises against continuation of this research line because of the low efficacy of this class of contraceptive.

1959

Seymour Katsh undertakes a study funded by the Population Council into the possibility of reviving research on immunological birth control methods for the 'control of populations'.

Late 1960s, early 1970s

A second wave of antifertility 'vaccine' research begins, conducted primarily by Pran Talwar, at that time of the All-India Institute of Medical Science (later director of the National Institute of Immunology, the NII); the population Council; and the Task Force on Vaccines for Fertility Regulation of the WHO's Special Programme of Research Development and Research Training in Human Reproduction (HRP).

1976–78

Human trials of Talwar's first beta-hCG immuno-contraceptive take place in five countries – India, Finland, Sweden, Chile and Brazil – under the auspices of the Population Council's International Committee for Contraceptive Research. Insufficient response is reported in one-quarter of the 63 (sterilized) women.

1976

Controversy develops over ethics after Talwar immunizes six fertile women without adequate animal trials and two women become pregnant. The WHO withdraws support and Talwar is forced to do more animal studies.

1987

Brazilian women's movement prevents trials by the Population Council.

1988

HRP completes Phase I (safety) trial of Vernon Stevens's anti-CPT beta-hCG prototype in Australia under Warren Jones. Some 10 of the initial 30 women on the trial were replaced after they complained of muscle and joint pains, attributed to rapid adjuvant release caused by the instability of the 'vaccine' emulsion.

Late 1980s

A number of citizen groups raise concerns about the development of antifertility immuno-contraceptives.

June 1989

HRP Symposium on Assessing the Safety and Efficacy of Vaccines to Regulate Fertility takes place; its major focus is the future of research on anti-hCG 'vaccines' (see Appendix 1).

November 1989

CONRAD and HRP co-sponsor the International Workshop on Gamete Interaction: Prospects for Immunocontraception (primarily on anti-egg and anti-sperm immuno-contraceptives). David Hamilton expresses doubts about the feasibility of safe immunization against body constituents.

1990

The NII completes its Phase I (safety) study of the new anti-beta-hCG immuno-contraceptive in 101 women. Problems of efficacy still exist and a decision is made to further change the contraceptive's formulation before going to phase II (efficacy) trials.

November 1990

The WEMOS Women and Pharmaceuticals Group calls HRP's and NII's Phase II trials 'premature', claiming that neither product offers 'significant advantages over already existing methods'. The Declaration of the Sixth International Women's Health Meeting in Manila asks for research on antifertility 'vaccines' to be stopped.

Ulrike Schaz documents the enrolment procedure for NII's phase II trials in a major New Delhi hospital. (Their film *Antibodies against Pregnancy* is launched in July 1991.)

February 1991

Women's health advocates at a meeting co-sponsored by HRP and the

International Women's Health Coalition, and entitled 'Creating Common Ground' ask HRP to convoke a special meeting on contraceptive vaccines.

The Population Council completes a Phase I trial of its anti-beta-hCG contraceptive in 24 women from Chile, Finland, and the Dominican Republic. Because of the greatly variable, often short-lived immune response, researchers conclude that an additional adjuvant may be needed. To date no further tests have been undertaken – partially because of the US anti-abortion lobby against funding the development of methods that act after conception.

August 1991

A Rio de Janeiro meeting of the Feminist International Network of Resistance to Reproductive and Genetic Engineering (FINRRAGE) asks for a stop to research.

1992

Talwar tests anti-GnRH immunization in 20 women who have just given birth.

August 1992

The HRP-organized meeting between Women's Health Advocates and Scientists to Review the Current Status of the Development of Fertility Regulating Vaccines takes place in Geneva. Most of the women's health advocates who attend express concern over the development of immunological birth control methods and ask for more information to enable them to discuss the issue in their home countries.

October 1992

Steven Sinding, director of population sciences at the Rockefeller Foundation, speaking in New Delhi on the fortieth anniversary of the International Planned Parenthood Federation, advises family planning policy makers 'to replace [demographic] program targets with objectives expressed in terms of stated desires of the people served', and draws attention to the link between population policies, coercion, and people's resistance to disrespectful family planning services.

March 1993

HRP co-sponsored International Symposium on Contraceptive Research and Development for the Year 2000 and Beyond is held in Mexico. Adrienne Germain, vice-president of the International Women's Health Coalition, recommends that 'no further work be done on the contraceptive vaccine'. Carmen Barroso, director of the Population Program of the MacArthur Foundation, calls on policy makers to face the 'implications' of dialogues with women's health advocates, namely 'the difficult process of shifting resource allocations'. Mahmoud Fatallah, senior adviser for biomedical health research at the Rockefeller Foundation and former director of the World Health Organization's HRP, calls for a 'second contraceptive technology revolution' away from a 'demographic-driven' to a 'women-centred' approach – as the best strategy to 'save the planet'.

June 1993

Start of the International Campaign to Stop the Research on Antifertility 'Vaccines' (Immunological Contraceptives). The campaign had been recommended by participants of the International Action Workshop co-organized by BUKO Pharma-Kampagne, Women's Health Action Foundation (WHAF) and Women's Global Network for Reproductive Rights (WGNRR) after the Bielefeld international conference entitled 'Antifertility "Vaccines": Researchers' Dream – Women's Nightmare?'

August 1993

The NII's Phase II (efficacy) trials studying an altered anti-hCG prototype in 148 women are completed. The product used in the Phase II trials contained a doubly cross-reacting antigen (alpha chain of sheep-LH coupled to beta-hCG) and used an 'alternate carrier' strategy (that is, for the three injections of the primary immunization schedule, the first carrier is a tetanus toxoid, the second diphtheria toxoid, and the third tetanus again), yet it is effective for only 80 per cent of the women. Average contraceptive duration is 3 months (results published in August 1994).

September 1993

Seventh International Women's Health Meeting takes place in Kampala, Uganda. The launch of *Vaccination against Pregnancy: Miracle or Menace?* (English-language version).

8 November 1993

A petition entitled Call for a Stop of Research on Antifertility 'Vaccines' (which called also for a radical redirection of contraceptive research) is sent to major research and funding organizations. Press and other public activities are organized by women's and health action groups in numerous countries. Talwar initially attributes international resistance to a few feminists and the professional jealousy of HRP.

January 1994

A. S. Paintal, a former director-general of the Indian Council of Medical Research, asks the Indian authorities to terminate, or at least suspend, any further trials with the NII's cross-reacting anti-hCG product.

The Rio Statement of the Reproductive Health and Justice International Women's Health Conference, co-organized by the International Women's Health Coalition and the Brazilian Citizenship Studies Information and Action Organization (CEPIA), recommends a redirection of 'resources ... from provider-controlled and potentially high-risk methods, like the vaccine, to barrier methods ... female-controlled methods that provide both contraception and protection from sexually transmitted diseases including HIV, as well as male methods'.

February 1994

Rumours about laced tetanus vaccines emerge in Tanzania. A Catholic priest alleges that tetanus vaccines given to schoolgirls and young women would cause abortion and sterility, and an article by Lawrence Roberge, of Human Life International, calls on churches to work against the anti-hCG 'abortion vaccine'. At around the same time, similar rumours are emerging in Peru, Mexico, the Philippines and Nicaragua. Attendances at immunization programmes decrease.

May 1994

The World Health Assembly, the annual meeting of the World Health Organization's policy-making body, takes place in Geneva. The organizations Antigena and Espace Femmes International stage a street theatre performance on immuno-contraceptives there; Health Action International and WHAF distribute a leaflet 'Immunological Contraceptives: Destined for Population not People'.

June 1994

Discussion on Ethical Aspects of Research, Development and Introduction of Fertility Regulating Methods, organized by HRP in conjunction with its Scientific and Ethical Review Group, takes place in Geneva. Its major focus is the abuse of contraceptives and criteria for research priorities.

HRP's Phase II (efficacy) trials, which had aimed to enrol 200–250 women in Sweden, are suspended. All of the first seven women volunteers had experienced moderate to strong pain in the injection area, in some cases radiating down the legs. Two, moreover, had developed sterile abscesses in the injection area, and four had fever. HRP suspected this was caused by the strong adjuvant (which they had added to offset the predictable low activity of their short hCG-antigen fraction).

September 1994

UN International Conference on Population and Development takes place in Cairo: antifertility 'vaccines' emerge as a recurrent theme in the NGO Forum. For example, DAWN launches a report on population and reproductive rights, which recommends an end to the development of antifertility 'vaccines'.

1995

The Indian Institute of Science's Phase I (safety) trial studying anti-FSH immuno-contraception in men is completed.

March 1995

The anti-abortion group Pro-Life Philippines wins a restraining order against further administration of tetanus vaccines alleged to contain an abortifacient.

June 1995

A number of participants at the 23rd Congress of the Medical Women's International Association express concern about the development of immunological birth control methods.

The Second International Action Workshop of the Campaign to Stop the Research on Antifertility 'Vaccines' takes place in Canada co-organized by Women's Health Interaction, Ottawa, the Committee on New Reproductive Technologies of the National Action Committee, Toronto, and the Women's Global Network for Reproductive Rights' Coordination Office, Amsterdam. Women attending the conference meet representatives of the International Development Research Council, a major co-funder of Talwar's research. The IDRC promises to increase the transparency of the research process, to involve women's groups in the review and monitoring process, and to issue a press statement correcting inaccuracies contained in a position paper it had released prior to the meeting.

The Dutch Minister of Development Cooperation, Jan Pronck, holds a second consultation meeting to discuss whether it should continue to fund research into immuno-contraceptives.

By now over 430 groups and organizations from 39 countries have signed the Call for a Stop of Research on Antifertility 'Vaccines'.

Call for a stop of research on antifertility 'vaccines' (immunological contraceptives)

We, the undersigned, call for an immediate halt to the development of immunological contraceptives because of concerns about health risks, potential for abuse, unethical research, and the assumptions underlying this direction of contraceptive research.

Groups of contraceptive researchers worldwide have been developing a completely new class of contraceptives for the past two decades. Immunological contraceptives, also known as antifertility 'vaccines', are being developed primarily for women in LACAAP[1] countries.

Five major institutions are currently carrying out the research:

— The National Institute of Immunology, New Delhi, India;
— The World Health Organization, Geneva, Switzerland;
— The Population Council, New York, USA;
— The Contraceptive Research and Development Program (CONRAD), Norfolk, USA; and
— The National Institute for Child Health and Development (NICHD), Bethesda, USA.

A variety of organizations are funding this research. They include the World Bank, the United Nations Population Fund (UNFPA), the United Nations Development Programme (UNDP), the Rockefeller Foundation, the US Agency for International Development (USAID), the International Research and Development Center (IDRC, Canada), and the governments of India, Norway, Sweden, United Kingdom and Germany.

The stated aim of those developing antifertility 'vaccines' is to induce temporary infertility by turning the immune system against body components which are essential for human reproduction. A variety of immunological contraceptives – mainly for women but also for men – are now being tested in clinical trials. The 'vaccine' which is most far advanced aims to neutralize the human pregnancy hormone hCG (human chorionic gonadotropin), a hormone produced in a woman's body by a fertilized egg shortly after conception. This hormone is

---

1. LACAAP = Latin American, Caribbean, Asian, African and Pacific

altered, then coupled to a bacterial or viral carrier (for example, a diphtheria or tetanus toxoid) so that the immune system mistakes the natural pregnancy hormone for an infectious germ and reacts against it. The body thus does not get a signal to prepare for pregnancy and the fertilized egg is expelled. Other immunological contraceptives are developed to interfere with the production of sperm, the maturation of egg cells, the fertilization process, or the implantation and development of the early embryo.

We oppose the development of immunological contraceptives. Our main reasons are:

1. *Abuse potential* Immunological contraceptives will not give women greater control over their fertility, but rather less. Immunological contraceptives have a higher abuse potential than any existing method. They will be long-acting (depending on the type they may last from one year to life-long). They cannot be 'switched off', and they are easy to administer on a mass scale because they will be injectables or a single pill. Researchers claim that the popularity and widespread acceptance of anti-disease vaccines could facilitate the introduction of antifertility 'vaccines', especially in LACAAP countries. This compounds our concern about mass administration of immunological contraceptives without people's knowledge or informed consent.

2. *Manipulation of the immune system for contraceptive purposes* Immunological contraceptives present no advantage for women over existing contraceptives. Because they use the immune system, they are inherently unreliable. Individuals can react completely differently to the same kind of immunological contraceptive. (For example, the Indian anti-hCG formula, the most advanced method, did not work for 20% of the women.) In addition, stress, malnutrition and disease will cause unpredictable failures of the contraceptive. In women and men with a predisposition to allergies and autoimmune diseases, on the other hand, the 'vaccine' may cause life-long sterility. People will have no outward sign to know whether and when an immunological contraceptive is working.

Immunological contraceptives are unlikely to ever be harmless. They interfere with delicate and complex immunological and reproductive processes. There are many potential risks: induction of autoimmune diseases and allergies, exacerbation of infectious diseases and immune disturbances, a high risk of fetal exposure to ongoing immune reactions. As research on antifertility 'vaccines' began 20 years ago, little or no thought has been given on how immunological contraceptives may directly or indirectly increase risks of HIV infection, or hasten the onset of full-blown AIDS.

Interference with the immune system for contraceptive purposes is indefensible at a time when primary health care systems in many countries are being dismantled, when the incidence of many infectious diseases is increasing, and when we have become acutely aware of the preciousness and complexity of our immune defense.

3. *Unethical clinical trials*   Clinical trials have taken place in India, Brazil, Sweden, Finland, Dominican Republic, Chile, Australia. Trials are currently taking place in Sweden (perhaps also in the United States) and further trials are planned in India.

International standards of ethics in clinical trials state that human experimentation should only take place if the product being developed offers advantages over existing options. Immunological contraceptives offer no advantage in terms of efficacy, reversibility, safety, protection against sexually transmitted diseases or control by the user. The risks to women and men cannot be justified.

In addition, these trials are of concern because:

— There has been insufficient testing of the anti-hCG contraceptive on animals before testing in humans; the animals used do not give enough indication of adverse effects for women and their children;
— The enrolment of women was not based on informed consent. The efficacy and safety of the immunological contraceptive have been overstated. Consent forms have compared immunological contraceptives to anti-disease vaccines. This analogy obscures the differences in principle of action and purpose between the manipulation of the immune system for contraception and the induction of immune defense against harmful micro-organisms;
— There has been insufficient data collection about adverse effects to women and to children born to women during trials.

4. *Framework of contraceptive research*   The interplay between population control institutions, Northern and Southern countries, religious institutions, the medical establishment, as well as the community, the society and male partners profoundly influences the type of contraceptives available and the way they are being provided – or not provided. As many researchers readily admit, the concept of antifertility 'vaccines' was conceived in a 'demographic driven, science led' framework. The major funders of contraceptive research want to increase the effectiveness of population control programmes. Most of the scientists involved in the research have been taught to see the body as a machine. The major trend in contraceptive development has been to create technologies which are long acting, have a low user failure rate, which lend themselves for mass fertility control – and which interfere with delicate and complex processes in the human body. Research on intrauterine devices, long-

acting hormonal injectables and implants (Norplant®) has been given precedence over user controlled low-tech methods such as diaphragms and condoms, or over existing local practices of fertility control. Women are seen as better contraceptive acceptors than men. Most contraceptive research is still directed at methods for women.

Antifertility 'vaccines' are a logical culmination of this framework.

What do we want?   We call for a radical reorientation of contraceptive research. Population control ideology should not guide the development of contraceptives. Approximately 10% of public funding for new contraceptive research worldwide is currently being spent on antifertility 'vaccines'. We would like to see this funding redirected. The aim must be to enable people – particularly women – to exert greater control over their fertility without sacrificing their integrity, health and well being. Contraceptive development must be oriented at the realities of women's lives. Above all it must consider local health care conditions and the position of women in society.

We call on all research institutions involved, in particular the National Institute of Immunology, the Population Council and the World Health Organization to immediately stop all research on immunological contraceptives.

We call on all funders to stop financing this type of contraceptive.

Author's note

The text reprinted above is an updated version of the petition published in November 1994.

List of signatories to the call for a stop of research on antifertility 'vaccines' (immunological contraceptives)

On November 8 1993 the petition was sent to the major research institutes and funding bodies involved in the development of immunological birth control methods. By June 1995 it had been signed by 434 groups and organizations from 39 countries, and by 250 individuals from countries represented in the list below and from Ethiopia, Laos, Pakistan, Tunisia and Uganda.

Australia

FINRRAGE, Parkville; The Women's Health Service for the West, Footscray; Women's Abortion Action Campaign, St James, NSW

Austria

Bund demokratischer Frauen Kärnten, Klagenfurt; Egalia, Verein für Fraueninitiativen, Vienna; Frauensolidarität, Vienna; Friedenswerkstatt, Linz; Österreichische Gesellschaft für Kritische Geographie, Vienna; Östereichischer Informationsdienst für Entwicklungspolitik (ÖIE), Vienna

Bangladesh

Gonoshastraya Kendra, Dhaka; Resistance Network, Dhaka; UBINIG, Dhaka; Women for Women, Dhaka

Barbados

WAND, Women and Development Unit, St Michael

Brazil

AABV, Boa Vista; ABEN-MA, São Luiz; Açao Cívica dos Cegos, Rio de Janeiro; ALERJ, Rio de Janeiro; AMDIRN, Natal; Apoio UNICA, Olinda; Assembléia Legislativa do Rio de Janeiro; Assistência Judiciária de OAB/RN, Natal; Associação Afro Brasileira Ogbon, São Paulo; Associação Brasileira de Mulheres da Carreira Jurídica, ABMCJ, Natal; Ass. Liberdade Mulher, Rio de Janeiro; CACES, Rio de Janeiro; Casa da Mulher do Nordeste, Recife; Casa da Mulher Grajaú; Casa do Parto, Centro Saúde Santa Rosa, Belo Horizonte; Casa Menina Mulher da CPP, Recife; CCSA-Depto. de Serviço Social, Maceió; CCSO-Núcleo de Estudios e Pesquisa sobre Condição Feminina, Maceió; CEAC, Rio de Janeiro; CEAP, Rio de Janeiro;

CEMINA, Niterói; CENPIA, Niterói; Centro Acotirano de Formação Popular, Maceió; Centro das Mulheres da Vila Brejal, Maceió; Centro Jos XXIII, Rio de Janeiro; Centro Solano Trinidade, Documentaçar e Pesquisa, Recife; CEPIA, Rio de Janeiro; CERP, Rio de Janeiro; CESAI/SESAB; CESTEH/FIOCRUZ, Rio de Janeiro; CIEPH, Florianópolis; Clube de Maes de Paratibe, Paulista; CMB – Laranjeiras, Rio de Janeiro; Colectivo Feminista Sexualidade e Saúde, São Paulo; Colectivo Mulher Vida, Olinda; Colectivo Refazendo, Recife; Conselho de Moradores do Planeta dos Macacos, Recife; Cons. Municipal dos Direitos da Mulher e das Minorias, Natal; COPPE/ UFRJ, Rio de Janeiro; CRIOLA, Rio de Janeiro; CSAU/UFAL; Cunha Colectivo Feminista, João Pessoa; CURUMIN, Recife; DEAF – COMPP-DOP, Rio de Janeiro; EMBRATEL, Rio de Janeiro; ENSP, Rio de Janeiro; ESPACO, São Paulo; FASE, Brasil; Federaçon das Mulheres do RN, Natal; FNETS – Tijuca, Rio de Janeiro; Forum de Mulheres de Pernambuce, Recife; Forum Mulheres Niterói; FURTAC – AC, Rio Branco; Geledes – Instituto da Mulher Negra, São Paulo; Grupo Autônomo de Mulheres, GAM, Natal; Grupo Dar á Luz, Brasília; Grupo de Gestantes, Rio de Janeiro; Grupo de Mulheres da Ilha, São Luis; Grupo de Mulheres da Sha, São Luis; Grupo de Parto Alternativo-Unicamp, Campinas; Grupo Justiça Seja Feita, Recife; Grupo Mulher Maravilha, Recife; IBGE, Rio de Janeiro; IDAC, Rio de Janeiro; INAMPS, Rio de Janeiro; Instituto Aurora de Yoga, Rio de Janeiro; ISER, Rio de Janeiro; MCBC, São Luiz; Metropolitana da Cidade de Recife; MMT Movimento Mulheres Trabalhadores, Paraíba; Movimiento Assistencial de Porto Alegre; Movimento de Mulheres em SC, Florianópolis; MUSA; Núcleo da Saúde da Mulher, Goiania; Oficina de Papel Meninos e Meninas do Recife; Ordem dos advogados do Brasil, Rio de Janeiro; Prefeitura Muniopal de Campinas; Promotora de Justica, Rio de Janeiro; REDEH Red de Defensa de Especie Humana, Rio de Janeiro; Secretaria da Mulher PT, Bayeux; SEEB/MA, São Luis; SENAI, Rio de Janeiro; SENUN, Rio de Janeiro; Sindicato das Parteiras Obstetras e Enfermeiras, Rio de Janeiro; Sindicato dos Trabalhadores Domésticos ma Area; Sindicato Trab. em Entidades Sindicais, Maceió; SINDSPREV, Niterói; SINTEP, João Pessoa; SOS CORPO, Grupo de Saude da Mulher, Recife; UFRJ – Geo Ciências, Rio de Janeiro; UNICAMP, Campinas; Universidade Federal da Paraíba; Universidade Federal de Santa Catarina, Florianópolis; VIA TV Mulher, Rio de Janeiro; VO.SE.F. Bo-Italía, Recife

Canada

Alternative Pour Elles, women's refuge; Centre de Femmes L'ERIGE, La Sarre; Conseil d'intervention pour l'accès des femmes au travail, Montréal; DES Action Canada, Montréal; Fédération du Québec pour le planning des naissances, Montréal; Femmes en Mouvement, Bonaventura; Inter Pares, Ottawa; Le centre des femmes La Sentin'Elle, Iles de la Madeleine; Le Collectif de Sept-Iles pour la Santé des Femmes; Le Collectif féministe de Rouyn-Noranda pour la santé des femmes; Maison Mikana, Amos; Naissance Renaissance Inc., Montréal; Regroupement des Femmes de Gaspé

Inc.; Regroupement des Centres de Femmes du Québec, Montréal; South Asian Women's Community Centre, Montréal; Transformative Learning Centre, Toronto; Vancouver Women's Health Collective; Women's Health Interaction, Ottawa

Chile

Casa de la Mujer La Morada, Santiago; Casa de la Mujer Valparaíso; Colectivo Arcilla, Santiago; Colectivo Atención Primaria de Salud, Santiago; Colectivo El Telar, Santiago; COMUSAMS, Colectivo Mujer Salud y Medicina Social, Santiago; EPES, Educación Popular en Salud, Santiago; FEMPRESS Mujer-Ilet, Santiago; Foro Abierto de Salud y Derechos Reproductivos, Santiago; Movimiento de Emancipación de la Mujer, Santiago

Denmark

KULU Women and Development, Copenhagen

El Salvador

DIGNAS, San Salvador

Fiji

Fiji Women's Rights Movement, Suva

Finland

Health Action International – Helsinki

France

MFPF, Mouvement Français pour le Planning Familial, Brignoles; Women Living Under Muslim Laws, Grabels

Germany

Agisra, women against sex tourism and trafficking in women, Frankfurt; AK Brasilien, Aachen; Aktionsgemeinschaft Solidarische Welt, Berlin; ALASEI – Agencia Latinoamericana de Servicios Especiales de Informacion, Bonn; Allerweltshaus Medienstelle, Köln; Apotheke am Viktoriaplatz, Berlin; Arbeitsgemeinschaft Katholischer Studenten und Hochschulgemeinden, Bonn; Arbeitskreis Dritte Welt, Oldenburg; Autonomer Frauenprojektbereich Gesamthochschule, Paderborn; Autonomes Referat für Frauen- und Lesbenpolitik, Duisburg; Berliner Frauengruppe gegen Bevölkerungspolitik; Bremer Informationszentrum für Menschenrechte und Entwicklung, Bremen; BUKO Pharma-Kampagne, Bielefeld; CARA, Bremen; Dritle Welt Forum, Aachen; Dritle Welt Laden, Kiel; Dritte Welt Haus, Bielefeld; Eskapaden, Feministische Bildungs- und Kulturarbeit, Hildesheim; Feministisches Frauenbildungswerk, Bielefeld; Feministisches Frauengesundheitszentrum, Berlin; Feministisches Frauengesundheitszentrum, Nürnberg; FINRRAGE, Feminist International Resistance against Reproductive and Genetic Engineering, International Coordination Office, Hamburg; Frauenbeauftragte der Universität, Hildesheim; Frauenbildungswerk eV, Bielefeld; Frauenbuchladen, Bielefeld; Frauenbüro, Darmstadt; Frauengleichstellungsbüro der Universität, Hildesheim; Frauengruppe gegen den §218, Freiburg; Frauen helfen Frauen, Frauenhaus, Bielefeld; Frauenkino, Bielefeld; Frauenkulturzentrum, Bielefeld; Frauen lernen gemeinsam, Bremen; FrauenLesben gegen Bevölkerungspolitik, Darmstadt; Frauenrat der Universität, Hildesheim; Frauen Selbsthilfe Laden, Hamburg; Frauen und Mädchen Gesundheitszentrum, Freiburg; Impatientia/Genarchiv, archive on genetic engineering and reproductive technologies, Essen; Katholische Student Innengemeinde, Münster; Kirchlicher Entwicklungsdienst (KED), Nürnberg; Mädchenhaus, Bielefeld; Mädchentreff, Bielefeld; MedizinstudentInnen, Würzburg; Notruf für vergewaltigte Frauen und Mädchen, Kassel; Nut – Frauen in Naturwissenschaft und Technik, Berlin; §218 Gruppe, Münster; Psychologische Frauenberatung, Bielefeld; Selbstverteidigung für Frauen eV, Bielefeld; SPD Ortsverein Münster-Coerde; Terre des Femmes, Tübingen; Terre des Hommes BRD, Osnabrück; VDPP – Verein Demokratischer Pharmazeutinnen und Pharmazeuten, Hamburg; Verein zur Unterstützung feministischer Mädchenarbeit, Bielefeld; Weltgruppe der evangelischen Kirchengemeinde Aplerbeck, Dortmund; Wildwasser, Bielefeld

Ghana

Society of Ghana Women Medical and Dental Practitioners, Korlebu

Great Britain

The Ecologist, Dorset

Haiti

ENFOFANM, Port au Prince

India

Aalochana, Pune; Action India, New Delhi; Adarsh Rural Integrated Development Society, Bukkapatnam; AIDS Awareness Group, New Delhi; AIDS Bhedbhav Virodhi Andolan; All India Democratic Women's Association; All India Drug Action Network, New Delhi; All India Mahila Dakshata Samiti, New Delhi; Alternative for India Development, Madras; Ankur, New Delhi; Ashadeep Vikas Kendra, Bombay; Association for Social Health in India, Bangalore; Bandra East Community Centre, Bombay; BEST, Bharat Environment Seva Team, Pudukkottai; Bhavatiya Mahila Federation, Nagpur; Centre for Women's Development Studies, New Delhi; Cheshive Home, Bombay; Chatra Yuva Sangharsh Vahini, Bombay; Chetana, Ahmedabad; Committee for Protection of Democratic Rights, Bombay; Community Health Cell, Bangalore; CREST, Bangalore; Dakshini Mahila Mandali, Bangalore; Deccan Development Society, Pastapur; Forum Against Oppression of Women, Bombay; Forum for Women's Health, Bombay; Hunda Virodhi Chalval, Beed; Indian People's Human Rights Commission, Bombay; Institute of Indian Culture, Bombay; Institute of Social Studies Trust, New Delhi; Institutional Association for Children's Development, Bombay; IWID, Madras; Jagori, New Delhi; Jagruthi Kendra, Bombay; Jagruthi Mahila

Sangh, Bangalore; Jeevan Nirvah Niketan, Bombay; JNM, Hyderabad; Join Women's Programme, New Delhi; Kali for Women, New Delhi; Karnataka Women Teachers Association, Bangalore; Khedut Mazdoor Chetna Sangath, Alirajpur; Lok Vignyan Sanghatana, Pune; Madras Institute of Development Studies; Mahila Sanskrutik Mand, Nagpur; Mahila Shakti, Tirupati; Majoor Sangharsh Vahini, Dhule; Manini, Bangalore; Multiple Action Research Group, New Delhi; Nari Samata Manch, Pune; Narmada Bachao Andolan, Bombay; National Federation of Indian Women, New Delhi; People's Union for Democratic Rights, New Delhi; Pragati Sheel Mahilla Samiti, New Delhi; Public Interest Research Group, New Delhi; Purogami Mahila Sangathan, Bombay; Purogami Mahila Sangathan, New Delhi; Purush Uvach, Pune; Pushpak FRIENDS NETWORK, Madurai; Roots Network, Pudukkottai; Sabla Sangh, New Delhi; Sabrang, Bombay; Safeline Madras; SAHELI Women's Resource Centre, New Delhi; Sahiyar, Baroda; Samaj Sewa Niketan, Bombay; Samajwadi Mahil Sabha, Dhule; Sewa Niketan, Bombay; Shakti Resource Centre for Women, Bombay; Shaktishalini, New Delhi; Shramik Kranti Sanghatana, Raigad; Shramik Mahila Morch, Pune; Shramik Mukti Dal, Dhule; Shramik Stri Mukti Sanghatana, Dhule; Shravanti Valley Development Society, Anakapalle; SNEHIDI, Madras; Speak India, Rapur; SRED, Society for Rural Education and Development, Arakonam; Stree Chetna, Nagpur; Stree Manch, Nagpur; Stree Mukti Sangathana, Bombay; Stree Shakti Sadan, Bombay; Stree Shakti Sadan, Ibave District; Stree Uvach, Bombay; Sumangali Seva Ashram, Bangalore; Tata Institute of Social Scientists, Bombay; UTTHAN Village Development Society, Patna; Vimo-chana, Bangalore; Women and Health Cell, Medico Friends Circle, Bombay; Women's Centre, Bombay; Women's Voice, Bangalore; YUVA, Bombay; YWCA, Bombay

Italy

Südtiroler HochschülerInnenschaft, Bozen

Japan

Karada Karada No Kai, Tokyo; Soshiren, Tokyo; Women's Centre Osaka

Kenya

MEDS, Mission for Essential Drugs and Supplies, Nairobi

Malaysia

Asian and Pacific Development Centre, Kuala Lumpur

Mali

OSEDA, Organization for Social and Environmental Development in Africa, Bamako

Mexico

CREFAL, Mexico City; GEM, Mexico City; Grupo EMAS, Mexico City; Hai Mexico, Mexico City; Mujeres de SEDEPAC, Mexico City; PIEM,

Programa Interdisciplinario de Estudios de la Mujer, Mexico City; PROESA, Mexico City; PSRS, Mexico City

Netherlands

CEBEMO, Catholic Organization for Joint Financing for Development Programmes, Oegstgeest; Coordination Office of Women's Global Network for Reproductive Rights, based in Amsterdam; COS Limburg, Maastricht; Department of Gender Studies in Agriculture, Wageningen; DES Action, Utrecht; Health Action International-Europe Coordinating Office, Amsterdam; HIVOS, Humanistic Institute for Cooperation with Developing Countries, The Hague; HOM, Humanistic Committee on Human Rights, Utrecht; ICCO, Interchurch Organization for Development Cooperation, Zeist; India Committee of the Netherlands, Utrecht; International Solidarity for Safe Contraception, Amsterdam; Netherlands Development Organization, Addis Ababa; Network of Women in Development Agencies in the Netherlands, Oegstgeest; OIKOS, Utrecht; SNV, The Hague; SNV Rwanda, Kigali; Stichting Grenzeloze Solidariteit, Eindhoven; Stichting Onderzoek & Voorlichting Bevolkingspolitiek, Heemstede; Tall Tree Consultancy, Amsterdam; Vrouwennetwerk, Zutphen; Wemos, Amsterdam; WHAF, Women's Health Action Foundation, Amsterdam; Women Mission Education, Valkenburg; Women's Group of the India Committee of the Netherlands, Utrecht; ISIS, Amsterdam

Nicaragua

Asociación de Mujeres Luisa Amanda Espinoza, Managua; Casa de la Mujer de Acahualinca, Managua; Casa de la Mujer Erlinda López – AMNLAE, Managua; Casa Materna de Matagalpa; Casa Materna de Ocotal Mary Barreda; Cátedra de Estudios de Género de la UCA, Managua; Católicas por el Derecho a Decidir, Managua; Centro de Mujeres de Masaya; Centro de Mujeres ISNIN, Managua; CEPRI Centro de Promoción y Rehabilitación Integral, Managua; CISAS, Managua; Centro de Adolescentes y Jóvenes de Sí Mujer, Managua; Clínica Xochilt Acalt, Léon; Colectiva de Mujeres de Masaya; Colectiva de Mujeres de Matagalpa; Colectiva La Malinche, Managua; Colectivo de Mujeres Itza, Managua; Colectivo de Mujeres 8 de Marzo, Managua; Colectivo de Mujeres Xochilt, Managua; Comisión Autónoma de Mujeres Ciegas, Managua; Fundación Xochiquetzal, Managua; Grupo de Mujeres Venancia, Matagalpa; IXCHEN, Centro de Mujeres, Managua; Ixim, Managua; Puntos de Encuentro, Managua; Red de Mujeres por la Salud 'María Cavalleri', Managua; Si Mujer Servicios Integrales Para la Mujer, Managua; Soy Nica, Managua

Nigeria

EMPARC, Empowerment Action and Research Centre, Lagos

Peru

CEM; Cendoc-Mujer, Lima; CESIP, Centro de Estudios Sociales y Publicaciones, Lima; CHIRAPAQ Centro de Culturas Indias, Lima; Flora Tristán, Lima; Movimiento Manuela Ramos, Lima

Philippines

BINHI Agricultural Resource Foundation, Inc., Quezon City; CWR, Centre for Women's Resources, Quezon City; Center for Environmental Concerns, Quezon City; Food For All Coalition, FFAC, Quezon City; GABRIELA Commission on Violence against Women, Quezon City; GABRIELA Commission on Women's Political Rights, Quezon City; GCWHRR, Quezon City; IAS, Institute for Alternative Studies, Manila; LAYA Women's Collective, Quezon City; MASAI, Quezon City; Mindanao Institute for Development, Inc.; NATL, Students Union of the Philippines – NUSP, Manila; Partnership in Development; WomanHealth, Quezon City

Puerto Rico

Grupo Pro Derechos Reproductivos, San Juan

Romania

Women's National Confederation, Bucharest

South Africa

Women's Health Project, Parktown; SAHSSO, South African Health and Social Services Organization, Booysens; St Scholasticas Clinic, Soekmekaar

Spain

Concertación Norte-Sur, Madrid; Esplai l'Albada, Reus

Switzerland

Antígena, Frauengruppe gegen Gentechnologie und Bevölkerungspolitik, Zürich; Arbeitsgruppe Schweiz–Kolumbien, Luzern; Arbeitskreis Tourismus und Entwicklung, Basel; ASDAC, Association suisse pour le droit à l'avortement et à la contraception, Bern; Basisgruppe Theologie, Fribourg; EcoSolidar, Zürich; Erklärung von Bern, Zürich; Espace Femmes International, Geneva; Feministisches Theologinnenforum, Fribourg; FIZ, Fraueninformationszentrum Dritte Welt, Zürich; FraP, Frauen machen Politik, Zürich; Frauen des Schweizerischen Friedensrates, Zürich; Frauen für den Frieden Aargau; Frauen für Aussenpolitik, deutsche Schweiz; Frauezitig FRAZ, Zürich; Genossenschaft Frauenambulatorium, Zürich; INCOMINDIOS Schweiz, Internationales Komitee für die indigenen Völker Amerikas, Basel; INFRA, Informationsstelle für Frauen, Zürich; MoZ, Gruppe Mutterschaft ohne Zwang, Zürich; NOGERETE, nationale feministische Organisation gegen Genund Reproduktionstechnologie, Zürich; Nottelefon für vergewaltigte Frauen, Zürich; OFRA, Organisation für die Sache der Frau, Basel; OFRA, Organisation für die Sache der Frau, Bern; OFRA, Organisation für die Sache der Frau, Zug; PRAKRITI INDIA, Schweizerischer Verein für die Erhaltung der Tier und Pflanzenvielfalt in Indien, Biel; Schwesterngemeinschaft, Zürich; Vereinigung der unabhängigen Ärztinnen und Ärzte der Region, Zürich; Vereinigung Dritte-Welt-Läden, Nd. Erlinsbach; Zentralamerika-Sekretariat, Zürich

Tanzania

KAMAJA, Kamati Ya Malezi Bora Ya Jamii, Ndanda; TAMWA, Tanzania Media Women's Association, Dar es Salaam

Thailand

Department of Community Pharmacy, Faculty of Pharmaceutical Sciences, Khon Kaen University; Friends of Women Foundation, Bangkok

Uruguay

Cotidiano Mujer, Montevideo

USA

Boston Women's Health Book Collective, West Somerville MA; Center for Democratic Renewal, Atlanta; Institute on Women and Technology, Amherst MA; Population and Development Program, Hampshire College, Amherst MA; Tufts University, Department of Urban and Environmental Policy, Medford MA; UBINIG, New York City NY; UCONN Students for Choice, Storrs CT

Zambia

Environment and Population Centre, Lusaka

Zimbabwe

Catholic Women's Clubs, Harare; SAPES, Southern Africa Political Economy Series, Harare; WAG, Women's Action Group, Harare; WASN, Women and Aids Support Network, Harare; WiLDAF, Women in Law and Development in Africa, Harare

Some useful organizations

Women's Global Network for Reproductive Rights

The Women's Global Network for Reproductive Rights is an autonomous network of groups and individuals in 114 countries. It was founded in 1978 to achieve and support reproductive rights for all women: their right to decide whether, when and with whom to have children. WGNRR strives for women's right to self-determination in keeping with their freedom, dignity and personally held values. Transforming social, political and economic conditions is part of this agenda. Among other things, network members have been working for women's unconditional access to: good-quality and holistic health services that respond to women's needs; comprehensive and unbiased information on sexuality and reproductive processes; fully informed, safe, effective and voluntary abortion; sexual self-determination without discrimination – and a world free of population policies and social norms that pressure some women into having children and other women into not having children. The co-ordinating office's work has been particularly focused on supporting the Campaign for the Prevention of Maternal Mortality and Morbidity, and, since 1993, on co-ordinating the International Campaign to Stop the Research on Antifertility 'Vaccines' (Immunological Contraceptives).

WGNRR, NZ.Voorburgwal 32, NL-1012 RZ Amsterdam, The Netherlands: phone +31-20-6209672; fax +31-20-6222450

BUKO Pharma-Kampagne

The Federal Congress of Development Action Groups (Bundeskongress entwicklungspolitischer Aktionsgruppen, BUKO) is a network of around 300 German solidarity groups. In 1989 BUKO started a campaign against global malpractices in drug marketing by multinational pharmaceutical companies. The focus of the Pharma-Kampagne is to stop unethical drug marketing practices and to foster rational use of drugs all over the world. The Pharma-Kampagne works with medical students, doctors, pharmacists and medical scientists,, through campaigns, publications, press work, public debate and dialogue. BUKO was one of the co-founders of HAI.

BUKO Pharma-Kampagne, August-Bebel-Str. 62, D-33602 Bielefeld, Germany: phone +49-521-60550; fax +49-521-63789

Health Action International

Health Action International (HAI) is an informal network of more than 150 consumer, health and development action and other public interest groups involved in health and pharmaceutical issues in 70 countries around the world. HAI actively promotes a more rational use of drugs through research, education, action campaigns, advocacy and dialogue. HAI has three co-ordination offices: Penang, Malaysia; Lima, Peru; and Amsterdam, the Netherlands.

HAI-Europe, Jacob van Lennepkade 334-T, NL-1053 NJ Amsterdam, The Netherlands: phone +31-20-6833684; fax: +31-20-6885002

HAI Clearinghouse and ARDA, c/o IOCU, PO Box 1045, Penang, Malaysia: phone +604-371396; fax +604-366506

AIS Latin America, c/o Accion para la Salud, Avda. Palermo 531, Dpto. 104, Lima, Peru: phone/fax +5114-7123202

Glossary

Adjuvant Any substance that enhances the immune response to the antigen with which it is mixed. Alum is the only adjuvant generally approved for humans.

Allergy Hypersentitivity to a specific *antigen*.

Antibody Proteins (called immunoglobins) produced by the body during an immune response. Antibodies are specific for one antigen and can build an antigen—antibody complex, thus neutralizing the antigen.

Antigen Any molecule that reacts with antibodies. However, some antigens do not by themselves elicit antibody production; only those antigens that can induce antibody production are called *immunogens*. Body tissues and substances can act as antigens. Yet usually they are not immunogenic because of the immune system's safeguards against reactions against healthy 'self' components, known as *immune-tolerance*. If immune tolerance fails, auto-immune disorders may ensue.

Antigenic determinant Most antigens react with several different antibodies. An antigenic determinant is the portion of an antigenic molecule bound by a given antibody.

Auto-immune disease A disease caused if the immune system reacts against specific healthy body cells or products as if they were harmful 'foreign' organisms. Auto-immune reactions can result in the destruction of tissue or interference with normal body functions.

Barrier methods Contraceptives that prevent pregnancy mechanically by preventing sperm from reaching the egg cell (condom, diaphragm, cervical cap).

Birth control 'vaccine' see *immuno-contraceptives*

Booster The restimulation of an immune response to a specific antigen through renewed contact with that original antigen.

Carrier A molecule that is linked to an antigen to turn it into an immunogen (see *antigen*). If the antigen is part of a molecule normally present in the body ('self' or 'self-like' antigen), then the carrier is needed to make it appear 'foreign' to the immune system. In the case of immuno-contraceptives, carriers include diphtheria, tetanus and cholera toxoids (the non-toxic version of those toxins); some teams now propose inserting the antigen into altered versions of viruses or bacteria (such as salmonella, vaccinia, fowlpox virus) through genetic engineering. The first type of 'vaccine' is called a carrier-based vaccine; the second a vector-based vaccine.

Contraceptive vaccine' see *immuno-contraceptives*

Cross-reaction Antibodies against a specific antigen sometimes react with other components that have a similar antigenic structure. Some auto-immune diseases are thought to be caused by antibodies against certain bacteria or viruses which also react against human body components that 'look' similar to them. In the case of the induction of immune reactions against human molecules it is difficult to predict which other body cells or structures will appear similar to the antibodies elicited and thus trigger auto-immune disease.

Depo Provera® Injectable contraceptive consisting of a synthetic form of the hormone progesterone (medroxyprogesterone acetate) which prevents conception for three months.

Dosage form The form of the completed pharmaceutical product. Dosage forms of hormonal contraceptives include tablets (the Pill), injectables and implants. Current immuno-contraceptive prototypes are mainly injectable products, but in the future some may be oral products.

Drug formulation The composition of a dosage form, including the characteristics of its raw materials (e.g. antigen-carrier conjugate, adjuvants, other substances constituting the solution, emulsion or tablet) and the operations required to process it.

Effectiveness The effectiveness of a contraceptive method is usually assessed by the number of unwanted conceptions per 1,200 cycles of use (i.e. per 100 'women years'). There is an important distinction between contraceptive effectiveness in controlled clinical study (theoretical or method effectiveness) and contraceptive effectiveness in general (actual effectiveness). In the case of immune contraceptives it seems important to distinguish between effectiveness in statistical terms and reliability for the individual user. Because of unpredictable variations of immune reactions the degree of effectiveness in percentage terms does not necessarily connote the degree to which individual women (or men) can rely on their contraceptive effect.

Failure rate The failure rate of a contraceptive method is usually expressed in terms of the percentage of women using a method who become pregnant accidentally during the first year of use. The failure rate can be seen as the inversion of the *effectiveness* rate (e.g. if a product has a method effectiveness rate of 95 per cent, then its method failure rate is 5 per cent). As for effectiveness rates, failure rates in well-controlled clinical studies, the method failure rate,, may be different from the actual failure rate of a contraceptive in the everyday life of the users.

FSH Follicle-stimulating hormone. FSH is released by the pituitary gland at the base of the brain, and in women stimulates (together with *LH*) the maturation of an egg cell in preparation for ovulation. In men, FSH stimulates the production and maturation of sperm.

GnRH Gonadotropin-releasing hormone. GnRH is released by the hypothalamus (part of the brain) and in turn stimulates the pituitary gland to release *FSH* and *LH*. These hormones are called gonadotropins because they act on the *gonads*.

Gonads Reproductive organs: the ovaries, which store egg cells in women, and the testes, which produce sperm in men.

hCG Human chorionic gonadotropin. A hormone released by the fertilized egg cell soon after fertilization, and later by the placenta throughout pregnancy. It stimulates the ovaries to maintain production of progesterone, a hormone that is necessary for the establishment and maintenance of a pregnancy.

Hormonal contraceptives Contraceptive methods based on synthetic derivatives of the hormones oestrogen and/or progesterone, which regulate women's menstrual cycles and other body functions. Hormonal contraceptives vary not only in terms of type and amount of synthetic hormones but also in terms of dosage form. Some are pills, others injectables or implants. Some *IUD*s now contain hormones (see *hormonal injectables*, *intra-uterine device* and *Norplant*®).

Hormonal injectables Hormonal contraceptives that have been formulated as oily injections in order to prolong the hormone release. Their intended action varies between one (Cyclofem), two (NET-EN) and three months (*Depo Provera*®). Until recently most injectables contained mainly synthetic derivatives of progesterone. Cyclofem contains both a synthetic oestrogen and a progesterone derivative.

Immune response The body's way of protecting itself against infection and foreign substances. A variety of specific and non-specific reactions are triggered after contact with an *antigen* (e.g. a virus, bacterium or toxin). A primary immune response follows the first contact with an antigen; secondary immune response follows second and further contact with the same antigen. Secondary immune responses involve a different type of antibody, have a shorter lag phase and are much more efficient.

Immune tolerance Failure of the immune system to react against a given antigen. Immune tolerance to our own body's healthy cells or products is an essential part of the immune system. If this 'self'-tolerance fails to function properly *auto-immune disease* can result.

Immunization Immunization or vaccination refers to procedures designed to increase an individual's level of immunity against a particular infectious agent or toxin. If this is achieved by administering an innocuous, altered form of the disease-causing virus, bacteria or toxin it is referred to as active immunization (because it relies on the active stimulation of the person's own immune system). In a way, active immunization mimics the immune stimulation caused by infectious organisms but aims to avoid the disease.

Immuno-contraceptives Contraceptive methods currently under development, which rely on an immune reaction against reproductive substances. Several types of immunological contraceptives are being researched for men and women. Synonyms: antifertility 'vaccines', birth control 'vaccines', immunological or immune-mediated contraceptives.

Immunogen Any *antigen* that elicits an immune response.

Implants Contraceptive implants; see *Norplant®*.

Intra-uterine device (IUD) A small plastic or metal device inserted into the uterus to prevent pregnancy. A variety of forms and types exist. Some IUDs are chemically inert, others release copper or a synthetic derivative of the hormone progesterone (the L-IUD).

IUD See *intra-uterine device*.

Lag period The period of time following immunization during which the immune response is built up to an effective level. The lag period after the initial 'primary' immunization is longer than that following subsequent 'booster' immunizations.

LH Luteinizing hormone. LH is a hormone secreted by the pituitary gland at the base of the brain; it stimulates ovulation in women and production of testosterone in men.

Norplant® A contraceptive consisting of six silicone capsules which are implanted under the skin of a woman's upper arm. They are filled with a synthetic form of the hormone progesterone (levonorgestrel), which is slowly released. They are effective for a period of five years. Minor surgery is required for insertion and removal of the capsules.

Self-tolerance see *immune tolerance*

Teratology The branch of sciene that deals with the abnormal development of the foetus and congenital malformations. Testing of any new pharmaceutical product has to include teratology trials in animals to evaluate whether any of its components may enter the uterus and harm the foetus.

Zona pellucida Translucent jelly-like membrane that surround the egg cell only in the later stage of cell maturation.

Bibliography

Ada, G. L. (1990) 'Immunosafety aspects of fertility control vaccines', in Alexander, N. J. et al. (eds) (1990), pp. 565–78.

Ada, G. L. and Griffin, P. D. (eds) (1991a) *Vaccines for Fertility Regulation: the assessment of their safety and efficacy*, World Health Organization, Cambridge University Press, Cambridge.

Ada, G. L. and Griffin, P. D. (1991b) 'Background and objectives', in Ada, G. L. and Griffin, P. D. (eds) (1991a), pp. 5–11 .

Ada, G. L. and Griffin, P. D. (1991c) 'The process of reproduction in humans: antigens for vaccine development', in Ada, G. L. and Griffin, P. D. (eds) (1991a), pp. 13–26.

Alan Guttmacher Institute (1992) *Norplant: opportunities and perils for low-income women*, Special Report 1, December.

Alexander, N. J. et al. (eds) (1990) *Gamete Interaction: prospects for immuno-contraception*. Proceedings of the Third Contraceptive Research and Development (CONRAD) Program International Workshop. Co-sponsored by the World Health Organization, San Carlos de Bariloche, Argentina, 26–9 November 1989, Wiley-Lyss, New York.

Anderson, D. J. and Alexander, N. J. (1983) 'A new look at antifertility vaccines', *Fertility and Sterility*, Vol. 40, No. 5, pp. 557–71.

'Anti-hCG vaccin: information om en klinisk prövning av en ny preventivmetod' (1993 (Anti-hCG vaccine: information about a new clinical trial of a new contraceptive method), at the Women's Health Clinics, Karolinska Hospital, Stockholm and University Hospital, Uppsala.

Australian Drug Evaluation Committee (1989) *Medicines in Pregnancy: an Australian categorisation of risks*, Australian Government Publishing Service, Canberra.

Bang, R. (1992) statement made at the 1992 HRP meeting to review the development of antifertility vaccines, Geneva, 17–18 August, on behalf of the Society For Education, Action and Research, Gadchiroli, India.

Bang, R. and Bang, A. (1992) 'Contraceptive technologies: experience of rural Indian women', *Manushi* 70, May/June, pp. 26–31.

Bardin, C. W. (1993) testimony before the United States House of Representatives, Committee on Small Business, Subcommittee on Regulation, Business Opportunities and Technology, New York, the Population Council, Center for Biomedical Research, 10 November.

Barriklow, D. (1993) 'An Indian institute's breakthrough vaccines', *Choices – The Human Development Magazine*, UNDP, 28–30 June.

Barroso, C. (1994) 'Meeting women's unmet needs: the allegiance between feminists and researchers', in Van Look et al., 1994, pp. 511–21.

Barzelatto, J. (1991) 'Welcoming address' in Ada, G. L. and Griffin, P. D. (eds) (1991a), pp. 1–3.

Basten, A. (1988) 'Birth control vaccines', *Ballière's Clinical Immunology and Allergy*, Vol. 2, No. 3, pp. 759–74.

Basten, A. et al. (1991) 'Assays for relevant human immune responses', in Ada, G. L. and Griffin, P. D. (eds) (1991a), pp. 75–93.

Beale, A. J., Research Division, Wellcome Biotechnology, Kent, UK (1991) communication at the 1989 HRP Symposium on Fertility Regulating Vaccines (see Ada and Griffin 1991a).

Benagiano, G. (1994) letter to Beatrijs Stermerding, WGNRR, 21 January.

Berelson, B. (1969) 'Beyond family planning: what further proposals have been made to "solve" the population problem, and how are they to be appraised', *Science*, Vol. 163, 7 February, pp. 533–43.

Berger, P. (1987) 'A cautionary view of antifertility vaccines', *Nature*, Vol. 326, p. 648.

BNF (*British National Formulary*) (1995), The Pharmaceutical Press, London.

Brache, V. et al. (1992) 'Whole beta-hCG: tetanus toxoid vaccine'. Paper presented at the Scientific Session on Immunological Aspects of Reproductive Health, on the Occasion of the 20th Anniversary of the WHO Special Programme of Research, Development and Research Training in Human Reproduction, Moscow, 16–18 June 1992.

Browne, M. W. (1991) 'New animal vaccines spread like diseases: in the wild, artificial viruses curb both rabies and rabbits', *New York Times*, 26 November 1991.

Bygdeman, Marc (1995) letter to Barbara Mintzes, Health Action International Europe Coordination Office, 24 May.

Catley-Carlson, M. (1994) letter to Beatrijs Stermerding, WGNRR, 28 June.

Chard, T. and Howell, R. J. S. (1991) 'Endocrinological hazards associated with human immunization with self or self-like antigens', in Ada, G. L. and Griffin, P. D. (eds) (1991a), pp. 95–120.

Chetley, A. (1990) *A Healthy Business: world health and the pharmaceutical industry*, Zed Books, London.

Chetley, A. (1995) *Problem Drugs*, Zed Books in association with Health Action International (HAI-Europe), London and Amsterdam.

CIOMS (1993) *International Ethical Guidelines for Biomedical Research involving Human Subjects*, prepared by the Council for International Organizations of Medical Sciences (CIOMS) in collaboration with the World Health Organization (WHO), Geneva.

Clarke, A. (1995) 'Negotiating the contraceptive quid pro quo c.1925–1963'. Paper presented at the Conference on Women, Gender and Science, University of Minnesota, 12–14 May 1995, to be published in A. Clarke (forthcoming), *Disciplining Reproduction: modernity, American life sciences and the problem of sex*, University of California.

Cohen, S. N. (1987) 'Immunization', in Stites, D. P. et al. (eds) *Basic and Clinical Immunology*, Appleton and Lange, Norwalk, Connecticut/Los Altos, CA, pp. 669–89.

Collingridge, D. (1980) *The Social Control of Technology*, Open University Press, Milton Keynes.

Concepcion, M. et al. (1991) 'Social aspects related to the introduction of a birth control vaccine', in Ada, G. L. and Griffin, P. D. (eds) (1991a), pp. 233–46.

Cookson, C. (1991) 'Vaccine is being developed as alternative to contraceptive pill', *Financial Times*, 18 August, p. 6.

Corea, G. (1991) 'Depo-Provera and the politics of knowledge', in Hynes, H. P. (ed.) *Reconstructing Babylon: essays on women and technology*, Indiana University Press: Bloomington, IN.

Corrêa, S. (1994) *Population and Reproductive Rights: feminist perspectives from the South*, Zed Books, London in association with Development Alternatives with Women for a New Era (DAWN).

Council for International Organizations of Medical Science (CIOMS) 1993 *International Ethical Guidelines for Biomedical Research Involving Human Subjects*, prepared by CIOMS in collaboration with the World Health Organization (WHO), Geneva, 63pp.

Davis, Angela (1990) 'Racism, birth control and reproductive rights', in Marlene Gerber Fried (ed.) (1990) *From Abortion to Reproductive Freedom*, South End Press, Boston, MA.

Declaration of People's Perspectives on 'Population' Symposium (1993), Comilla, Bangladesh, 12–15 December.

Duden, B. (1992) 'Population' in Sachs, W. (ed.) *The Development Dictionary*, Zed Books, London, pp. 146–57.

Dukes, G. (1995), letter to the author, 24 January.

Egjar, II. (1994) 'Vaccination against pregnancy – violence against women in the Third World', *Arbeidersbladet*, 31 May.

European Study Group (1989) 'Risk factors for male to female transmission of HIV', *British Medical Journal*, No. 289, pp. 411–15.

Fatallah, M. (1994) 'Fertility control technology: a women-centred approach to research', in Sen, G. et al. *Population Policies Reconsidered: Health, empowerment and rights*, Harvard University Press, Boston, MA, pp. 223–34.

FINRRAGE (Feminist International Network of Resistance to Reproductive and Genetic Engineering) (1991) 'Antipregnancy Vaccines, Position Paper', International Coordination Office, Hamburg, October.

Forum for Women's Health (1993) (formerly Forum Against Sex Determination and Sex Preselection), press release, Bombay, 30 October.

Fuente, M. de la (1995) 'Congress Medical Women's International Association', *WGNRR Newsletter* No. 50, April–June, pp. 33–4.

Furedi, F. (forthcoming), Polity Press, Cambridge.

Gabelnick, H. L. K. (1994) letter to Beatrijs Stemerding, WGNRR, 2 March.

Galazka, Arthur M. (1993) *The Immunological Basis for Immunization, Module 3: Tetanus*, Expanded Programme on Immunization (EPI), World Health Organization, Geneva.

Gaur, A. et al. (1990) 'Bypass by an alternate carrier of acquired unresponsiveness to hCG upon repeated immunization with tetanus conjugated vaccine', *International Immunology*, Vol. 2, No. 2, pp. 151–5.

Gelijns, A. C. (1991) *Innovation in Clinical Practice: the dynamics of medical technology development*, National Academy Press, Washington, DC.

Gevas, P. C. (1995) Comments addressing concerns about the WHO immunocontraceptive research programme, in fax to a women's health advocate, 31 May.

Gillespie, D. (1994) letter to Beatrijs Stemerding, WGNRR, 2 December.

Gomes dos Reis, A. R. (1988), 'WHO?', *Begleithefte zum zweiten Bundeskongres 'Frauen gegen Gen- und Reproduktionstechnologie'*, Bonn, pp. 10–12.

Gomes dos Reis, A. R. (1994) personal communication.

Gordon, L. (1990) *Women's Body, Women's Right: birth control in America* (revised and updated), Penguin Books, New York.

Griffin, P. D. (1987) 'A birth control vaccine: fertility control vaccines that would safely and effectively inhibit fertility without unacceptable side-effects would be attractive additions to the world's family planning armamentarium', *World Health*, November, pp. 25–7.

Griffin, P. D. (1988) 'Vaccines for fertility regulation', in HRP (1988), pp. 177–97.

Griffin, P. D. (1990a) 'Strategy of vaccine development', in Alexander, N. J. et al. (eds) (1990), pp. 501–22.

Griffin, P. D. (1990b) letter on contraceptive vaccine research, *Women's Global Network for Reproductive Rights Newsletter*, No. 32, pp. 6–7.

Griffin, P. D. (1992) 'Options for immunocontraception and issues to be addressed in the development of birth control vaccines', *Scandinavian Journal of Immunology*, Supplement, Vol. 36, No. 11, pp. 111–17.

Griffin, P. D. (1993) 'Fertility regulating vaccines: a background paper', Annex 2 in HRP (1993), pp. 37–44.

Griffin, P. D. (1994) letter to Beatrijs Stemerding, WGNRR (in response to a letter to the Swedish ethical committees responsible for HRP's Phase II trial), May 13.

Griffin, P. D. and Hendrickx, A. G. (1991) 'Animal models for assessing the safety and efficacy of antifertility vaccines', in Ada, G. L. and Griffin, P. D. (eds) (1991a), pp. 27–41.

Griffin P. D. and Jones, W. R. (1991) 'The preliminary clinical evaluation of the safety and efficacy of a fertility regulating vaccine', *Statistics in Medicine*, Vol. 10, pp. 177–90.

Griffin, P. D., Jones, W. and Stevens, V. (1994) 'Antifertility vaccines: current status and implications for family planning programmes', *Reproductive Health Matters*, No. 3, pp. 108–13.

HAI/WHAF (1994) *Immunological Contraceptives: designed for populations, not for people*, Health Action International-Europe and Women's Health Action Foundation, Amsterdam.

Hanhart, J. (1993) 'Women's views on Norplant: a study from Lombok, Indonesia', in Mintzes, B. et al. (eds) (1993), pp. 27–45.

Haraway, D. J. (1991) 'The biopolitics of postmodern bodies: constitutions of self in immune system discourse', in Haraway, D. J. (1991) *Simians, Cyborgs, and Women: the reinvention of nature*, Free Association Books, London.

Harding, S. (1991) *Whose Science? Whose Knowledge? Thinking from Women's Lives*, Cornell University Press, Ithaca, NY and Open University Press, Buckingham.

Hardon, A. (1989) 'An analysis of research on new contraceptive hCG vaccines', *Women's Global Network for Reproductive Rights Newsletter*, April–June 1989, pp. 15–16.

Hari, P. (1994) 'Vaccine controversy', *Business World*, 26 Jan.–8 Feb., pp. 96, 98.

Hartmann, B. (1991) 'Population Council report on Norplant in Indonesia', *Women's Global Network for Reproductive Rights Newsletter*, January–March 1991, pp. 21–4.

Hartmann, B. (1995) *Reproductive Rights and Wrongs: the global politics of population control*, South End Press, Boston, MA.

Health Alert (1994) 'Anti-contraceptive rumors hamper vaccination programme', 1–15 April, p. 1.

Heim, S. and Schaz, U. (1994) *Population Explosion: the making of a vision*, FINRRAGE Coordinating Office, Hamburg. (Introductiory essay published in *WGNRR Newsletter*, No. 45, January–March, pp. 9–10.)

Henderson C. J. et al. (1987) 'Contraceptive potential of antibodies to the zona pellucida', *Journal of Reproductive Fertility*, Vol. 83, No. 1, pp. 325–43.

Hope, J. (1992) 'A new era of fertility control', *Organon's Magazine on Women and Health*, No. 2, pp. 26–9.

HRP (1978) Task Force on Immunological Methods for Fertility Regulation, 'Evaluating the safety and efficacy of placental antigen vaccines for fertility regulation', *Clinical and Experimental Immunology*, Vol. 33, pp. 360–75.

HRP (1988) (Special Programme of Research, Development and Research Training in Human Reproduction) *Research in Human Reproduction: biennial report 1986–1987*, World Health Organization, Geneva.

HRP (1990) (Special Programme of Research, Development and Research Training in Human Reproduction) *Research in Human Reproduction: biennial report 1988–1989.* World Health Organization, Geneva.

HRP (1991) (Special Programme of Research, Development and Research Training in Human Reproduction) *Annual Technical Report 1990.*

HRP (1992) (Special Programme of Research, Development and Research Training in Human Reproduction) *Reproductive Health: a Key to a Brighter Future: biennial report 1990–1991.* Special 29th anniversary issue. World Health Organization, Geneva.

HRP (1993) *Fertility regulating vaccines: report of a meeting between women's health advocates and scientists to review the current status of the development of fertility regulating vaccines,* World Health Organization, Geneva, (Doc. WHO/HRP/WHO/93.1).

HRP (1994a) *Challenges in Reproductive Health Research: biennial report 1992–1993,* World Health Organization, Geneva.

HRP (1994b) 'Anti-hCG vaccine Phase II clinical trial status report', *Progress in Human Reproduction,* No. 30, p. 8.

HRP and IWHC (1991) *Creating common ground: women's perspectives on the selection and introduction of fertility regulating technologies.* Report of a meeting between women's health advocates and scientists, organized by HRP and the International Women's Health Coalition. Geneva, 20–22 February 1991.

HRP and SERG (1994) *Discussion on Ethical Aspects of Research, Development and Introduction of Fertility Regulating Methods,* Summary Report, Scientific and Ethical Review Group, HRP, Geneva, 1–2 June.

Hubbard, R. (1990) *The Politics of Women's Biology,* Rutgers University Press, New Brunswick, NJ.

Hubbard, R. and Wald, E. (1993) *Exploding the Gene Myth: how genetic information is produced and manipulated by scientists, physicians, employers, insurance companies, educators and law enforcers,* Beacon Press, Boston, MA.

IDRC (International Development Research Centre) (1995a) '*IDRC's Position*', Ottawa.

IDRC (International Development Research Centre) (1995b) '*Most Commonly Asked Questions on the Contraceptive Vaccine*', Ottawa.

IDRC (International Development Research Centre) (1995c) '*IDRC's Work in Women's Health Issues – An Overview*', Ottawa.

IDRC (International Development Research Centre) (1995d) '*Most Commonly Asked Questions on the Contraceptive Vaccine*' (2nd version), Ottawa.

IWHC and CEPIA (1994), report on the conference 'Reproductive Health and Justice: International Women's Health Conference for Cairo 1994', 24–28 January 1994, Rio de Janeiro.

Jacobson, J. L. (1991) *Women's Reproductive Health: the silent emergency,* Worldwatch Paper No. 102, Worldwatch Institute, Washington, DC.

Jayaraman, K. S. (1986) 'Contraceptive vaccines: trials in India launched', *Nature,* Vol. 323, p. 661.

Jones, W. R. (1982) *Immunological Fertility Regulation,* Blackwell Scientific Publications, Oxford.

Jones, W. R. (1986) 'Phase I clinical trial of an anti-hCG contraceptive vaccine', in Clark, D. A. and Croy, B. A. (eds) *Reproductive Immunology,* Elsevier, Amsterdam, pp. 184–7.

Jones, W. R. and Beale, A. J. (1989) 'Clinical parameters in pre- and post-registration assessment of vaccine safety and efficacy'. Paper presented at the Symposium on

Assessing the Safety and Efficacy of Vaccines to Regulate Fertility, Geneva, June 1989.

Jones, W. R. and Beale, A. J. (1991) 'Clinical parameters in pre- and post-registration assessment of vaccine safety and efficacy', in Ada, G. L. and Griffin, P. D. (eds) (1991a), pp. 147–64.

Jones, W. R. et al. (1988) 'Phase I clinical trial of a World Health Organization birth control vaccine', *Lancet*, 11 June 1988, pp. 1,295–8.

Joras, M. (1992) 'Vaccin contraceptif: bientôt les premiers essais cliniques chez la femme en France', *Le Quotidien du Médecin*, 24 January 1992.

Kanigel, R. (1987) 'Birth control: the vaccine alternative', *Span*, August 1987, pp. 26–9.

Katsh, S. (1959) 'Immunology, fertility and infertility: a historical survey', *American Journal of Obstetrics and Gynecology*, Vol. 77, No. 5, pp. 946–56.

LaCheen, C. (1986) 'Population control and the pharmaceutical industry', in McDonnell, K. (ed.) *Adverse effects: Women and the Pharmaceutical Industry*, IOCU, Penang, pp. 89–136.

Law, M. (1995) letter to Karen Seabrooke, Inter Pares, 28 March.

Levene, A. (1993) 'New contraceptive "vaccine" raises ethical questions', feature article, Reuters Amsterdam, 1 December.

Lewin, T. (1994) 'Dream contraceptive nightmare', *New York Times*, 8 July.

Loten, J. (1993) 'Gender and population: breaking the discipline barrier towards a multidisciplinary approach to sustainability'. Paper presented at the Seminar on Population, Gender and Sustainable Livelihoods, organized by the Society for International Development (SID) and the Sustainable Development Policy Institute (SDPI), sponsored by the International Development Research Institute (IDRC) in Islamabad, Pakistan.

McNally, R. (1994) 'Genetic madness: the European rabies eradication programme', *Ecologist*, Vol. 24, No. 6.

Marcelo, A. B. and Germain, A. (1994) 'Women's perspectives on fertility regulating methods and services', in Van Look, P. F. A. and Pérez-Palacios, G. (1994), pp. 325–42.

Martin, E. (1987) *The Woman in the Body: a cultural analysis of reproduction*, Beacon Press, Boston, MA.

Martin, E. (1994) *Flexible Bodies: tracking immunity in American culture – from the days of polio to the age of Aids*, Beacon Press, Boston, MA.

Mastroianni, L. et al. (eds) (1990) *Developing New Contraceptives: obstacles and opportunities*, National Academy Press, Washington, DC.

Mathur, A. (1993) 'No jobs, no food, but contraceptives, contraceptives', *Sunday Observer*, India, 11–17 July.

Mauck, C. P. and Thau, R. B. (1990) 'Safety of antifertility vaccines', *Current Opinion in Immunology*, No. 2, pp. 728–32.

Mehta, K. (1993) interview with Ulrike Schaz, December.

Mintzes, B. et al. (eds) (1993) *Norplant: Under her Skin*, Women's Health Action Foundation and WEMOS, Amsterdam.

Mitchison, N. A. (1990a), 'Gonadotrophin vaccines', *Current Opinion in Immunology*, No. 2, pp. 725–7.

Mitchison, N. A. (1990b) 'Lessons learned and future needs', in Alexander, N. J. et al. (eds) (1990), pp. 607–14.

Mitchison, N. A. (1991) 'Chairman's summary: present status and future prospects of antifertility vaccines', in Ada, G. L. and Griffin, P. D. (eds) (1991a), pp. 247–50.

Mitchison, N. A. (1993) 'Will we survive?', *Scientific American*, Vol. 269, No. 3 (special issue), pp. 102–8.

Morgall, J. M. (1993) *Technology Assessment: a feminist perspective*, Temple University Press, Philadelphia, PA.

Morsy, S., (1993) 'Bodies of choice: Norplant® Trials in Egypt', in Mintzes et al. (1993) pp. 89–114.

Moudgal, N. R. et al. (1988) 'Development of a contraceptive vaccine for use by the human male: results of a feasibility study carried out in adult male bonnet monkeys', in Talwar, G. P. (ed.) (1988) *Contraception Research for Today and the Nineties – progress in birth control vaccines*, Springer, New York, pp. 253–8.

Mwenge (1994) 'Scientific truth: vaccine against tetanus for abortion', February.

Nair, S. and Schaz, U. (1995) letter to Margaret Catley-Carson, director of the Population Council, 18 May.

Naz, R. (1988) 'The fertilization antigen: applications in immunocontraception and infertility in humans', *American Journal of Reproductive Immunology and Microbiology*, 16, pp. 21–7.

Nieschlag, E. (1986) 'Reasons for abandoning immunization against FSH as approach to male fertility regulation', in Zatuchni, G. I. et al. (eds) *Male Contraception: advances and future prospects*, Harper and Row, Philadelphia, PA, p. 395.

Nossal, Gustav J. V. (1993) 'Life, death and the immune system', *Scientific American*, Vol. XX Special Issue, September, pp. 20–31.

O'Sullivan, S. (1987) 'The politics of sexuality and birth control', in O'Sullivan, S. (ed.) *Women's Health: a Spare Rib reader*, Pandora, London.

Oudshoorn, N. (1994) *Beyond the Natural Body: an archeology of sex hormones*, Routledge, London and New York.

Paranjape, S. and Shah, C. (1993) letter to the author, 27 June.

PATH (Programme for Appropriate Technology in Health) (1994) 'Enhancing the private sector's role in contraceptive research and development', in Van Look, P. F. A. and Pérez-Palacios, G. (1994), pp. 393–451.

Pharma-Brief (1993) 'Damit der Alptraum nie Wirklichkeit werde: internationale Konferenz über immunologische Verhütungsmittel, No. 4–5, pp. 1–4.

Philippines Organizing Committee (1991) 'Resolution on Reproductive Rights', *In Search of Balanced Perspectives and Global Solidarity for Women's Health and Reproductive Rights: Proceedings of the Sixth International Women and Health Meeting*, Quezon City, Philippines, 3–9 November 1990.

Playfair, J. H. L. (1989) *Immunology at a Glance*, 4th edn, Blackwell Scientific Publications, Oxford.

Political Environments (1995) 'Controversy and coercion in India', Issue 2.

Population Council (1994) 'Development of contraceptive vaccines', unpublished position paper enclosed in Catley-Carson, M. (1994).

Reed, J. (1983) *The Birth Control Movement: from private vice to public virtue*, Princeton University Press, Princeton, NJ.

Report of the Symposium (1991) 'Points to consider in the assessment of the safety and efficacy of vaccines to regulate fertility', in Ada, G. L. and Griffin, P. D. (eds) (1991a), pp. 251–98.

Richter, J. (1992) 'Contraception: research on antifertility vaccines priority or problem?' *Women's Global Network for Reproductive Rights Newsletter*, No. 39, pp. 13–18.

Richter, J. (1993) *Vaccination Against Pregnancy: Miracle or Menace?* Bielefeld, Germany: BUKO Pharma-Kampagne and Amsterdam: Health Action International.

Richter, J. (1994a) 'Beyond control: about anti-fertility 'vaccines', pregnancy epidemics and abuse', in Sen, G. and Snow, R. C. (eds) (1994), pp. 205–31.

Richter, J. (1994b) 'Antifertility "vaccines": a plea for an open debate on the prospects of research', *WGNRR Newsletter*, No. 46, April–June, pp. 3–5.

Richter, J. et al. (1993) 'Inserting Norplant at all costs? A case report of a Norplant training session in Thailand', in Mintzes, B. et al. (eds) (1993), pp. 81–8.

Roberge, Lawrence (1994) 'Abortifacient vaccine looms as a new threat', *Express*, 17–19 March, p. 19.

Rose, N. R. et al. (1988) 'Safety evaluation of hCG vaccine in primates: autoantibody production', in Talwar, G. P. (ed.) (1988) *Contraception Research for Today and the Nineties – progress in birth control vaccines*, Springer, New York, pp. 231–51.

Rose, N. R. et al. (1991) 'Immunological hazards associated with human immunization with self or self-like antigens', in Ada, G. L. and Griffin, P. D. (eds) (1991a), pp. 121–46.

Sangi, Augustine (1994) 'No contraceptive properties in tetanus toxoid vaccines', *Sunday News*, 17 April.

Schaz, U. and Schneider, I. (1991) *Antibodies Against Pregnancy: the dream of the perfect birth from the laboratory*. A documentary film by U. Schaz with the assistance of I. Schneider. English copy available through U. Schaz, Bleicherstr. 2, D-22767 Hamburg, Germany.

Schneider, I. (1991) 'Immun gegen Schwangerschaft', *Entwicklungspolitische Korrespondenz*, No. 3, September 1991, pp. 15–18.

Schrater, A. F. (1992) 'Contraceptive vaccines: promises and problems', in Bequaert Holmes, H. (ed.) *Issues in Reproductive Technology, Vol. I: An anthology*, Garland Publishing, New York, pp. 31–52.

Schrater, F. A. (1994a) 'Immunization to regulate fertility: biological concerns and questions', in Sen, G. and Snow, R. C. (eds), pp. 255–66.

Schrater, F. A. (1994b) 'The pros and cons: guarded optimism', *Reproductive Health Matters*, No. 4, November 1994, pp. 109–10.

Segal, S. J. (1991) 'The role of technology in population policy', *Populi*, Vol. 18, No. 4, pp. 5–13.

Segal, S. J. (1993) 'Trends in population and contraception', in *International Symposium on the Recent Advances in Female Reproductive Health Care, Proceedings*, Finnish Population and Family Welfare Federation, Helsinki, pp. 11–20.

Sen, G. and Snow, R. C. (eds) (1994) *Power and Decision: the social control of reproduction*, Harvard University Press, Boston, MA.

Senanayake, P. and Turner, J. (1994) 'Fertility regulation: history and current situation – an NGO perspective', in Van Look and Pérez-Palacios (1994), pp. 249–62.

Shearman, R. P. (1982) 'Foreword' in Jones, W. R. (1982), pp. vii–viii.

Sherwin, S. (1989) 'Feminist ethics and new reproductive technologies', in Overall, C. (ed.) The *Future of Human Reproduction*, Women's Press, Toronto, Ontario.

Sinding, S. W. (1994) 'Seeking common ground: unmet needs and demographic goals', *International Family Planning Perspectives*, Vol. 20, No. 1, pp. 23–7.

Snow, R. (1994) 'Each to her own: investigating women's response to contraception', in Sen, G. and Snow, R. C. (eds) *Power and Decision: the social control of reproduction*, pp. 233–53.

Spieler, J. (1987) 'Development of immunological methods for fertility regulation', *Bulletin of the World Health Organization*, Vol. 65, pp. 779–83.

Spieler, J. (1992) 'An overview of funding of vaccine research'. Unpublished paper presented at the meeting of HRP for women's health advocates, members of the

Steering Committee of the Task Force on Vaccines for Fertility Regulation, vaccine developers and representatives of interested agencies, Geneva, 17–18 August 1992.

Spierkel, K., Director IDRC Public Information Program (1995), letter to Karen Seabrooke, Inter Pares, Ottawa, 14 June.

Stackhouse, J. (1994) 'Magic bullet or just a shot in the dark?', *Globe and Mail*, Toronto, 9 March, pp. A1–A2.

Staines, N., Brostoff, J. and James, K. (1993) *Introducing Immunology*, 2nd edn, Mosby, St. Louis and London.

Stemerding, B. (1995) 'International action meeting to stop the research on antifertility "vaccines"', *WGNRR Newsletter*, No. 50, April–June, pp. 3–5.

Stevens, V. C. (1986) 'Current status of antifertility vaccines using gonadotropin immunogens', *Immunology Today*, Vol. 7, No. 12, pp. 369–74.

Stevens, V. C. (1990) 'Birth control vaccines and the immunological approaches to the therapy of non-infectious diseases', *Infectious Disease Clinics of North America*, Vol. 4, No. 2, pp. 343–54.

Stevens, V. C. (1992) 'Future perspectives in vaccine development', *Scandinavian Journal of Immunology*, No. 36, Supplement 11, pp. 37–143.

Stevens, V. C. et al. (1990) 'Studies of various delivery systems for a human chorionic gonadotropin vaccine', in Alexander, N. J. et al. (eds) (1990), pp. 549–64.

Sunny, S. and Shah, J. (1994) 'India's birth control vaccine: new hope or new hazard?', *Health and Nutrition*, January, pp. 20–27.

Suresh, N. (1994) 'Contraceptive vaccine project: ten years and still no sign of them', *Times of India*, 12 May.

Symbahayan Commission, Philippines (1995) 'Antifertility-drug experiments?', *Population Research Institute Review*, March–April, p. 9.

Talwar, G. P (1990) 'Reproduction: editorial overview', *Current Opinion in Immunology*, Vol. 2, pp. 723–4.

Talwar, G. P. (1994a) 'Advances in contraceptive vaccine research', *South to South Newsletter*, No. 3, June, pp. 1–3.

Talwar, G. P. (1994b) 'Immuno-contraception: guest editorial', *Current Opinion in Immunology*, No. 6, pp. 698–704.

Talwar, G. P. (n.d.) 'Current status and future of immunological approach to fertility control', National Institute of Immunology, New Delhi, 2pp.

Talwar, G. P. and Raghupathy, R. (1989) 'Antifertility vaccines', *Vaccine*, Vol. 7, April 1989, pp. 97–101.

Talwar, G. P. et al. (1990a) 'Phase I clinical trials with three formulations of anti-human chorionic gonadotropin vaccine', *Contraception*, Vol. 41, No. 3, pp. 301–16.

Talwar, G. P. et al. (1990b) 'Experiences of the anti-hCG vaccine of relevance to development of other birth control vaccines', in Alexander, N. J. et al. (eds) (1990), pp. 579–94.

Talwar, G. P. et al. (1990c) 'Two vaccines under clinical trials for control of fertility and reproductive hormone dependent cancers'. Paper presented at the Third International Symposium on Contraception, Heidelberg, Germany, 19–23 June 1990.

Talwar, G. P. et al. (1992a) 'Vaccines for control of fertility', paper presented at the HRP meeting between women's health advocates and scientists to review the current status of the development of fertility-regulating vaccines, Geneva, 17–18 August 1992 and the 8th International Congress in Immunology, Budapest, Hungary, 23–26 August 1992.

Talwar, G. P. et al. (1992b) 'Vaccines for the control of fertility and hormone-dependent cancers', *International Journal of Immunopharmacology*, Vol. 14, No. 3, pp. 511–14.

Talwar, G. P. et al. (1993) 'A birth control vaccine is on the horizon for family planning', in Finnish Population and Family Welfare Federation (ed.) *International Symposium on the Recent Advances in Female Reproductive Health Care*, Proceedings 5–6 June 1992, Helsinki, pp. 115–20.

Talwar, G. P. et al. (1994a) 'A vaccine that prevents pregnancy in women', *Proc. Natl. Acad. Sci*, USA, Vol. 91, August, pp. 8, 532–6.

Tan, M. (1995) 'Anti-tetanus vaccines and abortion', *Health Alert*, Special Issue, 1–15 May, pp. 2–6.

Task Force on Vaccines for Fertility Regulation, HRP (1978) 'Evaluating the safety and efficacy of placental antigen vaccines for fertility regulation', *Clinical and Experimental Immunology*, Vol. 33, pp. 360–75.

Thau, R. (1992) 'The Population Council's anti-fertility vaccine research'. Paper presented at the meeting of HRP, Geneva, 17–18 August 1992.

Thau, R. (1993) personal communication with Nicolien Wieringa, Women's Health Action Foundation, February.

Thau, R. et al. (1987) 'Long-term immunization against the beta-subunit of ovine luteinizing hormone has no adverse effects on pituitary function in rhesus monkeys', *American Journal of Reproductive Immunology and Microbiology*, No. 15, pp. 92–8.

Thau, R. et al. (1990) 'Advances in the development of antifertility vaccine', in Mettler, L. and Billington, W. D. (eds) *Reproductive Immunology*, Elsevier, Amsterdam, pp. 237–44.

Times of India (1994) 'Birth control vaccine: unethical tests on women alleged', 5 January.

Townsend, S. (1990) 'Norplant: safe and highly effective', *Network*, Vol. 11, No. 4, December 1990, Family Health International.

Tripathy, S. (1979) 'Anti-pregnancy vaccine in trouble', *India Express*, 5 May 1979.

UBINIG (1990) 'Norplant, the five year needle: an investigation of the Norplant trial in Bangladesh from the user's perspective', *Issues in Reproductive and Genetic Engineering*, 1990, No. 3, pp. 211–28.

Van Look, P. F. A. and Pérez-Palacios, G. (1994) *Contraceptive Research and Development 1984 to 1994: the road from Mexico City to Cairo and beyond*, Oxford University Press on behalf of the World Health Organization, New Delhi .

Vigy, M. (1992) 'Les essais hâtifs du "vaccin" contraceptif', *Le Figaro*, 25 January 1992.

Vines, G. (1994) 'Time to throw away your old contraceptives?' *New Scientist*, Vol. 142, No. 1,923, 30 April, pp. 36–40.

Wajcman, J. (1991) *Feminism Confronts Technology*, Polity Press, Cambridge.

Wajcman, J. (1994) 'Delivered into men's hands? The gender relations of reproductive technologies', in Sen, G. and Snow, R. C. (eds), pp. 153–75.

Ward, S. J. et al. (1990) 'Service delivery systems and quality of care in the implementation of Norplant in Indonesia'. Report prepared for the Population Council, New York, February 1990.

Weber, J. W. C. (1991) 'Regulatory aspects of vaccine development', in Ada and Griffin (1991a), pp. 185–200.

Wellin, S. (1993) 'Some issues in research ethics', in Wellin, S. (ed.) *Scientific Responsibility and Public Control, Studies in Research Ethics*, No. 2, Royal Society of Arts and Sciences in Gothenburg.

WGNRR (1993) 'Population and development policies: report on the international conference "Reinforcing Reproductive Rights"', *WGNRR Newsletter* (special insert) No. 43, April–June.

WHO (World Health Organization) (1986) 'First human trial of birth control vaccine begins in Australia'. WHO press release, Geneva, 17 February 1986.

WHO (World Health Organization) (1992) 'AIDS: a community commitment. Biggest world AIDS day ever'. WHO press release, Geneva, 20 November 1992.

Wieringa, N. (1994) 'Antifertility vaccines: the wrong road?', editorial, *International Journal of Risk and Safety in Medicine*, No. 6, pp. 1–9.

Wieringa, N. (1995) letter to Marge Berer and Sundari Ravidran, co-editors of *Reproductive Health Matters*, 12 January.

Will, A. (1995) letter to Margaret Catley-Carson, director of the Population Council, 15 May.

World Medical Association Declaration of Helsinki: recommendations guiding physicians in biomedical research involving human subjects, adopted by the 18th World Medical Assembly in Helsinki in 1964, last amended by the 41st World Medical Assembly in Hong Kong 1989, in e.g. CIOMS (1993) pp. 47–50.

Index